ARTIFICIAL CELLS

Publication Number 828

AMERICAN LECTURE SERIES®

A Monograph in

The BANNERSTONE DIVISION *of*
AMERICAN LECTURES IN LIVING CHEMISTRY

Edited by

I. NEWTON KUGELMASS, M.D., Ph.D., Sc.D.
Consultant to the Department of Health and Hospitals,
New York, New York

ARTIFICIAL CELLS

By

THOMAS MING SWI CHANG, B.Sc., M.D., C.M., Ph.D., F.R.C.P.(C).

Professor of Physiology
Faculty of Medicine
McGill University
Montreal, Canada

Associate
Medical Research Council of Canada

CHARLES   C   THOMAS   •   PUBLISHER

Springfield   •   Illinois   •   U.S.A.

Published and Distributed Throughout the World by

CHARLES C THOMAS • PUBLISHER

Bannerstone House

301-327 East Lawrence Avenue, Springfield, Illinois, U.S.A.

Natchez Plantation House

735 North Atlantic Boulevard, Fort Lauderdale, Florida, U.S.A.

© 1972, by CHARLES C THOMAS • PUBLISHER

ISBN 0-398-02257-7

Library of Congress Catalog Card Number: 77–169875

With THOMAS BOOKS *careful attention is given to all details of manufacturing and design. It is the Publisher's desire to present books that are satisfactory as to their physical qualities and artistic possibilities and appropriate for their particular use.* THOMAS BOOKS *will be true to those laws of quality that assure a good name and good will.*

Printed in the United States of America

PP-22

To

Lancy,
Harvey, Victor, Christine, Sandra
and my parents

FOREWORD

O UR Living Chemistry Series was conceived by Editor and Publisher to advance the newer knowledge of chemical medicine in the cause of clinical practice. The interdepedence of chemistry and medicine is so great that physicians are turning to chemistry, and chemists to medicine in order to understand the underlying basis of life processes in health and disease. Once chemical truths, proofs, and convictions become sound foundations for clinical phenomena, key hybrid investigators clarify the bewildering panorama of biochemical progress for application in everyday practice, stimulation of experimental research, and extension of postgraduate instruction. Each of our monographs thus unravels the chemical mechanisms and clinical management of many diseases that have remained relatively static in the minds of medical men for three thousand years. Our new Series is charged with the *nisus élan* of chemical wisdom, supreme in choice of international authors, optimal in standards of chemical scholarship, provocative in imagination for experimental research, comprehensive in discussions of scientific medicine, and authoritative in chemical perspective of human disorders.

Dr. Chang of Montreal presents his own investigations and the studies from other laboratories as fresh starting points for the possible synthesis of the smallest unit of the human body capable of independent function, the cell. It is the basic structural unit of all living matter, the true locus of health and disease. Physical systems of cellular dimensions described as models of artificial cells, theoretically and experimentally, are devised to supplement or replace defective or deficient cells in the body. Each must provide the minimal equipment of at least three units: first, a system of membranes for enclosing the cell, compartmentalizing the interior, controlling the chemical economy, and packaging the key catalysts of the cell; second, an apparatus for reproducing exact copies of the cell and for replicating key parts; and third, a mechanism for powering the activities of the cell through coupled oxidations. As yet, the exact form, content, and location of this equipment within the cell are difficult to delineate and to reconstitute. The living cell within a selective and retentive membrane contains a complete set of the different kinds of units necessary and sufficient to permit its own growth and reproduction from simple nutrients. Cell activities are based on a complex interplay, not of molecules alone, but of molecules organized into cellular subunits or organelles. It is the interplay of these substructures, mediated through the intracellular flux of molecules, that determines the behavior of the cell.

The line which separates the living from the nonliving has become somewhat obscured by recent developments in the life sciences. Contemporary investigators are hard put to say whether or not a crystallized virus is a living organism or whether or not a strand of nucleic acid reproducing itself in a test tube is alive. The current creed of the scientist is that all phenomena which characterize life processes can be described in chemical and physical terms and that the principles of chemistry and physics which apply to the inanimate world are equally valid for the world of life. The cell is more than a container for the vital machinery. It is veritably an expression of a universal set of mechanistic principles and of a unique molecular architecture and structural pattern. There may be alternative forms of life on other planets, based on a unit other than the cell, but the only unit of life we know is the cell. Cell scientists are as important as their discoveries in teaching us the spirit of exploration in the effectiveness of their techniques. There are the "achievers" who apply established methods, the "creators" who work independently with their own methodology and the "problem-solvers" who formulate cell problems for experimental solution. They are like leaves, each putting for the leaf that is created in them. Their imagination imitates but their critical spirit creates.

> "In creating, the only hard thing is to begin;
> A grass blade's no easier to make than an oak."

I. Newton Kugelmass, M.D., Ph.D., Sc.D., *Editor*

PREFACE

SOME biomedical research workers are more interested in investigating the clinical implications of basic advances in chemical, physical, engineering, and medical sciences. Back in 1956, I was surprised to find that despite the fundamental importance of cells, serious attempts had not been made to use the available basic knowledge in chemistry and biomedical sciences to investigate the feasibility of preparing "artificial cells." With the naiveness of an uninitiated, I proceeded to look into this in a corner of the physiology teaching laboratory at McGill University, equiped with a borrowed clinical centrifuge, a few chemicals, and a few glasswares. After a few frustrating months in 1957, I finally came up with samples of artificial cells, each consisting of hemoglobin and enzymes from red blood cells enveloped in an ultrathin spherical polymer membrane of cellular dimensions. *Artificial cell* is not a specific physical entity; it is an idea involving the preparation of artificial structures of cellular dimensions for possible replacement or supplement of deficient cell functions. Thus, different physical systems can be used to demonstrate this idea. Indeed, further studies in this laboratory and an increasing number of other laboratories have resulted in different physical systems, each demonstrating the feasibility of the idea of "artificial cell." One of the approaches developed in this laboratory is at present being tested clinically. The present monograph will concentrate on those systems which have been demonstrated experimentally to have potential in clinical applications.

The reasons for preparing this monograph are many. There is an increasing number of laboratories becoming interested in the general area of "artificial cells." From discussions and questions raised, it is clear that details of unpublished materials and updating of published materials from this laboratory can be rather useful for those working in this area. Furthermore, many of the publications on this topic from this and other laboratories have appeared in different specialty journals not readily available to investigators in unrelated specialties. In addition, this area of research is so interdisciplinary in nature that an attempt has been made to prepare this monograph in such a way that it can be useful to basic physical scientists, engineers, biomedical scientists, and clinicians, both as an introduction to this area of research, and also with sufficient up-to-date technical details for those who want to carry out research in this area. The only exceptions are the chapters on the preparation and the biophysical properties of the "artificial cell," where, of necessity, more specialized terminology and descriptions are required. The preparative procedures have been updated and simplified so that some of them can be reproduced by workers using only apparatus and chemicals readily available in any research laboratories, for example a magnetic stirrer, a clinical centrifuge, a few glasswares, and a few chemicals. More elaborate and complicated approaches are also described.

The first part of this monograph is a consideration of the general characteristics of artificial cells and specific theoretical examples of artificial cell systems. This is followed by examples of typical preparative procedures which have been updated since the original publications. The biophysical properties of artificial cells are characterized. They are then used in experiments designed to test some of the theoretical examples. Recent results of actual clinical application of one of the principles of artificial cells will also be included. From these experimental and clinical results, the implications of the general principle of "artificial cell" in biomedical research and in clinical therapy will be discussed. Thus, besides factual description of preparative procedures and experimental results, emphasis in this monograph will also be on the principles involved, possible lines of extension, and unexplored areas for further investigations.

T. M. S. Chang

ACKNOWLEDGMENTS

THIS research would not have been possible without making full use of the vast amount of information in the physical and biomedical sciences gathered by numerous investigators throughout the centuries. More immediately, there are many people who have at one time or another given invaluable encouragements, support, advice, or discussions. It will not be possible to acknowledge all, except to mention a few examples. In 1956 Professor F. C. MacIntosh, F.R.S., while Chairman of the Department of Physiology, was kind enough to let me try out my idea on "artificial cells" in the department and has, in addition, given me much valuable advice, and discussions. Professor A. V. Burgen, F.R.S., while at McGill, has through his teaching and discussions initiated my deep interest in the general area of membrane and cell physiology. In 1962, Professor S. G. Mason, F.R.S.C., was kind enough to offer me his laboratory facilities and advice for six months for me to improve and extend my preparative procedures. I am thankful to the Canadian biomedical scientists for their open enthusiasm in encouraging and supporting this study when it was first presented at the Canadian Federation of Biological Societies in 1963. My sincere appreciation goes to Professor W. Kolff, Professor A. Burton, Professor O. Denstedt, and many others for their interest, encouragement, and support; Professor D. V. Bates, F.R.C.P., Chairman of the Department of Physiology, for his support and help in improving the research facilities; and Professor J. Beck, F.R.C.P., Physician-in-Chief of the Royal Victoria Hospital, for his support in my clincial trials with Dr. A. Gonda, Dr. M. Levy and Dr. J Dirks. I acknowledge with thanks Mr. K. Holeczek for his earlier technical assistance and the preparation of the illustrations in this monograph; Mr. D. Cameron, chief technician of the Department of Physiology, for his earlier help; Mrs. N. Malave for carrying out the very tedious technical work since 1964, and Mr. C. Lister, Mrs. T. Lee Burns, Mrs. A. Verzosa, Miss E. Taroy, and others in the later years; past and present graduate students, medical students, and summer students, especially Dr. M. Poznansky, Mr. A. Rosenthal, Miss L. Johnson, Dr. A. Pont, Dr. O. Ransome, Mr. K. S. Lo, Mr. S. Chuang, Mr. B. Lesser, Mr. E. Siuchong, for participating in this research; and Mr. R. Enns for reading part of the manuscript. My wife, Lancy, through her cheerful acceptance has made it possible for me to combine this research with the heavy load of a medical student in the early phases, and later, in the last ten years, in addition to her moral support, she has helped with the typing and preparation of many of the publications, including typing and reading the rough copies and the final copy of this manuscript.

It is a pleasure to accept the invitation of Dr. I. N. Kugelmass and Charles C Thomas, Publisher, to prepare this monograph; their advice, patience, and understanding is deeply appreciated.

I wish to express my appreciation to the publishers of the following journals for permission to reproduce published materials and illustrations: *Canadian Journal of Physiology and Pharmacology; Transactions of the American Society of Artificial Internal Organs; Science; Nature; Journal of Biomedical Material Research; Science Journal; Science Tools; Federation Proceedings.*

I gratefully acknowledge the financial support of my parents in the first summer of 1957; the Faculty of Medicine at McGill for support throughout the next three summers; and, when this research was carried out on a full time basis since 1962, the Medical Research Council of Canada under the presidency of Professor G. Malcolm Brown, F.R.C.P., for the research support and for research appointments respectively as Medical Research Fellow (1962-1965), Medical Research Council Scholar (1965-1968), and presently, the career investigator appointment of Medical Research Council Associate (since 1968). The recent effort of the Medical Research Council in helping to develop this study for clinical trials is appreciated.

T. M. S. C.

CONTENTS

ARTIFICIAL
CELLS

Chapter 1

GENERAL CONSIDERATION

INTRODUCTION

WE all feel humble in the face of the ingenuity and complexity of nature. Yet, it need not prevent us from learning, exploring and even attempting to construct clinically useful artificial systems having a few of the simpler properties of their natural counterparts. Indeed, working on a molecular level, chemically oriented investigators have already synthesized a number of biological molecules, including those with the complexity of polypeptides and enzymes. On the level of organ and tissue substitutes, artificial kidneys, heart-lung machines, synthetic heart valves, vascular replacements, cardiac pacemakers, and others have become indispensable tools in the practice of medicine. However, on a cellular level, despite the fundamental importance of cells as unit structures of organs and tissues, the clinical potential of "artificial cells" has not been intensively investigated. This may have been due to the general feeling that such an endeavor would be premature, since our present basic knowledge of biological cells is still incomplete. Yet, organ substitutes like artificial kidneys and heart-lung machines are by no means exact replicas of their biological counterparts. Despite this, their clinical usefulness cannot be disputed. In the same way, to prepare clinically useful artificial cells, one may not have to reproduce exact replicas of biological cells. With this in mind, I started in 1956 to prepare and study simple artificial cell systems. Starting with a simple system, modifications and extensions have led to the experimental demonstration of possible clinical applications and the recent clinical use of one of these approaches.

BASIC FEATURES OF ARTIFICIAL CELLS

"Artificial cell" is not a specific physical entity. It is an idea involving the preparation of artificial structures of cellular di-

3

mensions for possible replacement or supplement of deficient cell functions. It is clear that different approaches can be used to demonstrate this idea. My initial attempts were to model the simplest of biological cells, red blood cells. Each of the artificial cells consists of a spherical ultrathin polymer membrane enveloping a microdroplet of hemoglobin and enzymes from hemolysate. The membranes of artificial cells were initially made of spherical ultrathin (500 Å) cellulose nitrate (Chang, 1957). Later, many other types of synthetic polymers such as Silastic® rubber (Chang, 1966) and those prepared from interfacial polymerization (Chang, 1964; Chang *et al.*, 1966) were used. To simulate biological cell membranes, biological materials were also used as the enclosing membranes: protein (Chang, 1964, 1965, 1969e; Chang *et al.*, 1966), polysaccharide (Chang *et al.*, 1967b), lipids (Mueller and Rudin, 1968; Pagano and Thompson, 1968), a complex of lipid and cross-linked protein or of lipid and polymer (Chang, 1969b). These later systems were attempts to mimic more closely the chemical compositions of biological cell membranes of a complex lipoprotein structure coated externally by mucopolysaccharides.

Artificial cells possess another feature of biological cells in that they are membrane-enclosed aqueous compartments of microscopic dimensions. In this form, besides having cell membranes which can be very thin without losing any mechanical strength, there is an enormous surface area. Since, among other factors, the rate of movement of permeant molecules across a membrane is proportional to its surface area and inversely proportional to its thickness, the rate of exchange across 1 ml of 20μ diameter artificial cells is many times faster than that across a single sphere with a volume of 1 ml. Thus, 1 ml of artificial cells (20μ in diameter) has a total surface area of 2,500 cm^2 (Chang, 1966); whereas, a single sphere 1 ml in volume has a surface area of only about 5 cm^2. Furthermore, 200 Å thick ulthrathin spherical membranes have sufficient mechanical strength to form stable membranes for the microscopic artificial cells, but much thicker membranes are required for the 1 ml sphere. In addition, the small diameter of artificial cells, like its biological counterpart, allows molecules, which equilibrate rapidly across this enormous surface area of very thin membranes, to diffuse quickly throughout the very small intracellular space. In biological cells, the advantage of a large surface-to-volume relationship is clearly demonstrated by the observation that smaller red blood cells have a greater velocity constant for the initial uptake of oxygen than do the larger ones (Holland and Forster, 1966). Experiments show that the large surface area of artificial cells also allows for extremely rapid equilibration of permeant solutes (Chang, 1966; Chang and Poznansky, 1968c).

Another general feature of biological cells has also been partially mimicked in artificial cells. Biological cell membranes, through passive restriction or special transport mechanisms, create an intracellular environment that differs from the extracellular. Thus, through passive restriction (Fig. 1), impermeant intracellular macromolecules and organelles (I) are conserved within both the biological and artificial cells and prevented from coming into direct contact with impermeant external materials (E). The enclosed impermeant materials (I) are prevented from escaping into the extracellular environment where they might be excreted or metabolized or might develop harmful effects. At the same time, by simple diffusion or special carrier mechanisms, permeant molecules (s,p) can equilibrate rapidly across the membranes to be acted upon by the cell content. For instance, the membranes of red blood cells and artificial cells do not allow the enclosed macromolecules (I) like hemoglobin or carbonic anhydrase to leak out. Yet, the enclosed hemoglobin (I) can act as an oxygen carrier by combining reversibly with extracellular oxygen (s,p). Carbonic anhydrase (I), while retained within both types of cells, acts enzymatically in converting carbon dioxide (s) to carbonic acid (p). Besides carbonic anhydrase, many other enzyme systems (I) of biological and artificial cells are also located intracellularly either in solution or in association with intracellular membranes. While confined within the cells, these enzyme systems act on permeant external molecules (s) entering both types of cells, converting substrates (s) into suitable products (p).

Manifold variations in the properties and the permeability characteristics of the enclosing artificial membranes are possible,

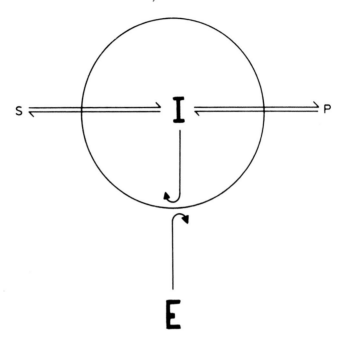

I   - Intracellular impermeant materials

E   - Extracellular impermeant materials

S,P - permeant molecules

Figure 1. Schematic representation of biological and artificial cells.

thus permitting the selective permeation of various types of molecules. These include variations in polymers, proteins, lipids, polysaccharides, charge, pores, thickness, and surface properties. Transport carrier mechanisms like Valinomycin have also been incorporated into artificial cell membranes (Chang, 1969b). The mean diameters and contents of these artificial cells can also be varied at will.

This general principle of artificial cells can form the basis of a large number of artificial systems of cellular dimension. Some of these might be briefly mentioned to illustrate the general principle. Later chapters describe experiments for testing the feasibility of these systems.

ARTIFICIAL CELLS CONTAINING CELL HOMOGENATE OR ENZYMES

My initial interest (Chang, 1957), was to place cell homogenate inside artificial cells (Figs. 1 and 2). This way, the biological cell membrane is replaced by an artificial membrane. The artificial membrane does not possess the heterogeneous or homologeous antigens present in the biological cell membrane. Furthermore, the enclosed impermeant cell content (I) of heterogeneous origins does not leak out to become involved in immunological reactions, and extracellular antibodies (E), if present, would not enter the membrane to affect the content. At the same time, permeant molecules (s,p) can cross the artificial membrane to interact with the content (I). This system was initially investigated using a model system in which red blood cell contents were enveloped in spherical ultrathin artificial membranes (Chang, 1957, 1964). This way, while permeable to oxygen and carbon dioxide, the artificial membrane would retain the enclosed hemoglobin and carbonic anhydrase intracellularly. The enclosed hemoglobin and carbonic anhydrase would thus be prevented from entering the extracellular environment and becoming involved in an immunological reaction, or being excreted or metabolized. Furthermore, the artificial membrane would lack the blood group antigens normally present in the erythrocyte membranes. In addition, the artificial polymer membranes would not become fragile after prolonged *in vitro* storage. It is obvious that this system of artificial cells containing cell homogenates, illustrated in the model system of enclosure of red blood cell hemolysate, can also be investigated for the enclosure of other cell contents, for instance, homogenate of liver and endocrine cells with their complex system of enzymes and intracellular organelles (Chang, 1965; Chang *et al.*, 1966).

A large array of enzyme systems is pesent in the cell homogenates enclosed in artificial cells. If an enzyme or enzyme system can be obtained in the synthetic, purified, or crude state, it can be added to supplement the original array of enzyme systems in the cell homogenate, or simpler artificial cells can be made to contain one enzyme system instead of the complex system of cell homogenate (Figs. 1 and 2) (Chang, 1964; Chang

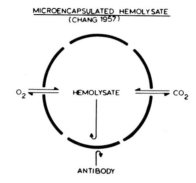

MICROENCAPSULATED HEMOLYSATE
(CHANG 1957)

O_2 ⇌ HEMOLYSATE ⇌ CO_2

ANTIBODY

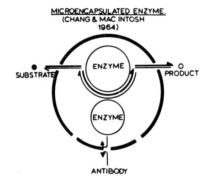

MICROENCAPSULATED ENZYME.
(CHANG & MAC INTOSH
1964)

SUBSTRATE ← ENZYME → O PRODUCT

ENZYME

ANTIBODY

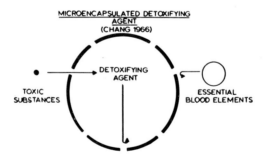

MICROENCAPSULATED DETOXIFYING
AGENT
(CHANG 1966)

DETOXIFYING
AGENT

TOXIC
SUBSTANCES

ESSENTIAL
BLOOD ELEMENTS

Figure 2. Schematic representation of three typical examples of artificial cell systems. *Upper,* artificial cells containing hemoglobin and enzymes from the contents of red blood cells. *Middle,* artificial cells containing enzymes. *Lower,* artificial cells containing detoxicant. (From Chang, 1966. Courtesy of American Society for Artificial Internal Organs.)

and MacIntosh, 1964b). The enclosed enzyme (I) would similarly be prevented from leaking out and from involvement in immunological reaction (E) and yet would be able to act freely on extracellular permeant substrate (s). This would avoid the problems associated with protein sensitization, anaphylactic reaction, or antibody production as a result of the repeated administration of a heterogeneous enzyme.

Certain groups of inborn errors in metabolism due to enzyme deficiency are candidates for investigation along this line (Chang, 1964, 1969d; Chang and MacIntosh, 1964b; Chang and Poznansky, 1968a). If the deficient enzyme has been synthesized, purified, or isolated, it can be easily enclosed in these artificial cells. However, in many enzyme deficiency diseases the deficient enzymes have not yet been purified or isolated. In these cases, if the cellular location of these enzyme systems is known, artificial cells containing the homogenate or extracts of these cells might be used. The most suitable cases for investigation would be those where the substrate (s) and its products (p) are molecules which can readily cross the membranes of the artificial cells. There are other conditions which might be investigated by means of artificial cells, even though the conditions are not due to the deficiency of a particular enzyme system. Examples of these conditions include the use of artificial cells containing asparaginase to lower the asparagine level in lymphosarcoma (Chang, 1969c, 1971a), urease to lower the urea level in uremia (Chang, 1966), and cholinesterase to trap the organic phosphate in poisoning. In addition, there is the possibility of investigating whether erythrocyte, hepatic, endocrine, or other organic and cellular deficiencies could be replaced by artificial cells containing the respective cell homogenates (Chang, 1965; Chang, *et al.*, 1966).

ARTIFICIAL CELLS CONTAINING INTACT CELLS

Transplantation of endocrine cells within millipore chambers has been attempted (Brooks, 1960a,b); however, these are always followed later by rejection of the graft. It has been suggested that these rejections are due to diffusion limitation leading to failure of nutrition of the enclosed cells (Brooks and Hill,

1960; Bassett, *et al.*, 1960; Stone and Kennedy, 1962). Restricted diffusion of nutrient may be present from the outset because of the thickness of the millipore filter (100μ-150μ thick) and the low surface-to-volume ratio (1.75 cm in diameter). Later, plugging of the filter by insoluble calcium salts and overgrowth of fibrous materials might further impede diffusion. Furthermore, millipore filters, though not permeable to leucocytes, are permeable to proteins, so that when antibodies are present in high concentration in the body fluids, the enclosed homografts are destroyed (Gabourel, 1961; Russell and Monaco, 1964). It has been suggested (Chang, 1965) that the enclosure of intact cells and cell cultures in artificial cells of suitable size might offer a solution to these difficulties. In the artificial cell system, the semipermeable membrane can protect the enclosed material from leucocytes and antibodies. In addition, biologically compatible membranes (Chang et al., 1967a,b) can be used to prevent fibrous coating of the membrane. At the same time, the large surface-to-volume ratio and the ultrathinness of the membrane would allow rapid diffusion of essential nutrient for the cells. Specific permeant cellular products of the enclosed intact cells would also enter the general extracellular compartment of the recipient. For instance, it would be interesting to see whether endocrine cells placed within artificial cells might survive and maintain an effective supply of hormone for the recipient (Chang, 1965). There would be the further advantage that implantation could be accomplished by a simple injection procedure rather than by a surgical operation. In addition, large-scale enclosure of intact cells might be investigated for replacement in organ deficiency or enzyme deficiency diseases. For example, a culture of liver cells may be enclosed in such artificial cells.

ARTIFICIAL CELLS CONTAINING BIOLOGICALLY ACTIVE SYNTHETIC MATERIALS

There are a number of biologically active synthetic materials which may be used *in vivo*. Some of these may require enclosure in an intracellular environment to permit selective action without any undesirable interaction with the environment.

Typical examples are ion exchange resins and adsorbents like activated charcoal. They can efficiently remove unwanted metabolites and toxins from blood perfusing over them. Unfortunately, many of these detoxifying agents also remove or adversely affect some dissolved or formed elements of blood, especially platelets and leucocytes. In addition, in the case of activated charcoal granules, there are problems involved with the embolism of fine powders. Using the general principle of artificial cells (Figs. 1 and 2), detoxifying agents (I) could be enclosed in artificial cells, with membranes having a number of special properties (Chang, 1966, 1969e). Membranes should not absorb or be permeable to essential elements of blood (E). They should be impermeable to the enclosed detoxifying agent (I), but at the same time freely permeable to external unwanted metabolites or toxins (s,p). This way, unwanted extracellular metabolites or toxins (s,p) would diffuse across the membrane of the artificial cells and be absorbed by the enclosed detoxicant (I). However, the artificial cell membrane would prevent the intracellular detoxicant from coming into direct contact with impermeant external material (E) like plasma protein, essential electrolytes, formed elements, and other essential dissolved elements of blood. This way, the enclosed detoxicant, while acting on permeant extracellular molecules (s,p) would be prevented from having any deteriorating effects on the essential elements of blood (E). In addition, the membrane can be made nonthrombogenic, so that no anticoagulants need to be added to the circulating blood (Chang *et al.*, 1967a,b). Furthermore, any fine powders escaping from particulate detoxicants like activated charcoal are prevented from embolising (Chang, 1969e; Chang and Malave, 1970). Many synthetic materials with other biological functions might also be enclosed in the artificial cells in a similar way to give them an intracellular environment.

Other biologically active synthetic materials which are biologically compatible may not require enclosure in spherical ultrathin membranes. Thus solid silicone rubber microspheres (Chang, 1966) and emulsion of fluorocarbon (Sloviter and Kaminoto, 1967; Geyer *et al.*, 1968) have been used to replace red blood cells. In these forms, although having no definite enclosing membranes, the silicone rubber or fluorocarbon material is im-

miscible with the extracellular environment. The materials them-
selves create an intracellular environment for oxygen and carbon
dioxide to diffuse into.

BASIC STUDIES IN MEMBRANE BIOPHYSICS

Many ingenious approaches have been used to study the biophysical and physiological properties of biological membranes.
Bungenberg de Jong and his co-workers used solid lipid coacervates (1935) and soap micelles (1956) to study permeability
phenomena. Schulman and Rideal (1937) used the monolayer
technique to great advantage in studying the physicochemical
aspects of membranes. Others used permselective membranes
(Sollner, 1968; Carr *et al.*, 1945) to study the effects of charged
membrane on the movement of electrolytes. Bangham and his
co-workers (1965) found that liquid crystals, each consisting of
concentric shells of bimolecular lipid layers, are very useful for
various permeability studies. Tobias' group (1964) coated Millipore filters with different types of lipids while Mueller and
Rudin (1962) started the very popular membrane model of
black lipid films consisting essentially of forming a bimolecular
lipid film in a small hole.

The artificial cells (Chang, 1957, 1964) described in this
monograph consist of microscopic spherical envelopes of ultrathin membrane separating an intracellular aqueous phase from an
external aqueous phase. Since the membrane compositions can
be varied at will, they can serve as additional models for the
study of biophysical properties of membranes. Indeed, work in
this laboratory shows that it is possible to vary the membrane
properties as to porosity, thickness, charge, lipid content, protein content, mucopolysaccharide content, and polymer composition. We have used artificial cells with different membrane
properties to study the biophysics of membrane transport
(Chang and Poznansky, 1968c), the relationship of surface
properties to survival in circulation (Chang, 1965; Chang *et al.*,
1967b), and the relationship of physicochemical properties to
the effect on coagulation and formed elements of blood (Chang
et al., 1967b, 1968, 1969). Other workers have made use of these
artificial cells for the study of the mechanical and electrical

properties of membranes (Jay, 1967; Jay and Edwards, 1968; Jay and Burton, 1969; Jay and Sivertz, 1969). Recently, Mueller and Rudin (1968) have adopted the procedure of the preparation of artificial cells (Chang, 1964) to prepare artificial cells having a bilayer lipid membrane. These are used in permeability studies. Pagano and Thompson (1967), Seufert (1970) and others have also prepared spherical ultrathin membranes of lipids for the study of movement of electrolytes. Since biological cell membranes consist of protein and lipids, we have recently used artificial cells with lipid-coated ultrathin spherical polymer or lipid-coated cross-linked protein membranes in membrane transport studies using macrocyclic molecules (Chang, 1969b).

CELL PHYSIOLOGY

In cell physiology, there is much research being done with the problems relating to the origin of cellular responses; for instance, the question of whether the triggering mechanisms are located at the cell surface, in intracellular organelles, or in the soluble constituents of the cytoplasm. The enclosure of cell contents in artificial cells might well provide preparations with which studies of this kind would become feasible (Chang, 1965). For instance, it would be interesting to test the sites of action of certain hormones; changes in extracellular electrolytes; and substances causing the release of certain intracellular material. This would give information as to whether the substance acts on specific surface receptors and only indirectly affects the intracellular material or whether it directly affects the intracellular material.

OTHER SYSTEMS

The general principle of artificial cells could be explored in other areas (Chang, 1965). Thus microencapsulation of radioactive isotopes or of antimetabolites might be used for intra-arterial injection into tumour-bearing tissue. In this case some of the microcapsules might lodge at the tumour site, while others would be carried by lymphatic channels to metastases in regional lymph nodes. Microencapsulation of radiopaque material would provide a contrast medium. Provided microcapsules can be made that will circulate readily in the bloodstream, they might be used as vehicles for contrast materials in angiography. The microencapsulation of highly magnetic alloys might provide a useful preparation for the measurement of blood flow in unopened vessels by electromagnetic techniques. The weak ferromagnetism of hemoglobin has been used successfully for this purpose, but instrumentation could be much simplified if microcapsules containing magnetic alloys could be made to circulate. If membranes of cross-linked protein can be made to retain the immunological characteristics of the protein, there might be a place for microcapsules in serological studies. Microencapsulation of drugs for slow release (Luzzi, 1970a) after oral or parenteral administration is another possibility.

Chapter 2

PREPARATION

GENERAL

$\mathbf{W}$HEN this study was started in 1956, a search of the published scientific literature showed that no methods had been reported for the preparation of aqueous suspensions of structures of cellular dimensions, each consisting of a spherical ultrathin semipermeable membrane enveloping an interior of biologically active material. Thus the first problem was to devise a method. My initial attempts used a spraying technique in which a powder or emulsion of red blood cell hemolysate was sprayed at right angles through a collodion spray (Chang, 1957). Later, I used another approach consisting of three main steps (Fig. 3) (Chang, 1957): first, the aqueous protein solution was emulsified in an organic liquid to form aqueous microdroplets; then, on the addition of a cellulose nitrate solution to the stirred emulsion, a membrane was formed on the surface of the microdroplets and allowed to set in butyl benzoate; and finally, the microcapsules were dispersed in an aqueous phase. With modifications (Chang, 1964; Chang *et al.*, 1966), this general procedure has been varied for the preparation of artificial cells with a great variation in membrane materials, content, and diameter. Obviously, it is not possible to describe in detail all the possible variations; instead this chapter describes the basic principles of preparation of artificial cells as illustrated by updated preparative procedures. Possible variations will be briefly described. A brief review will also be given describing the potentiality of industrial microencapsulation technology for use in the preparation of artificial cells.

It should again be emphasized that "artificial cell" is a concept. The artificial cells prepared by the following procedures are but physical examples for demonstrating this concept. There is no doubt that modification of the present system or completely

15

different systems can be made available to further demonstrate the feasibility of artificial cells.

ORGANIC PHASE SEPARATION
(Interfacial Precipitation)
Introduction

In this process the formation of the membrane depends on the lower solubility of a polymer, dissolved in a water-immiscible fluid, at the interface of each aqueous microdroplet in an emulsion (Chang, 1957). When suitable conditions are present, a number of polymers can be used this way to form the enclosing membranes of artificial cells. This principle is illustrated below, with cellulose nitrate as the membrane material and an erythrocyte hemolysate as the internal phase. The following procedure is updated and simplified from the one described earlier (Chang, 1957, 1964, 1965; Chang *et al.*, 1963, 1966).

Typical Example

The starting materials are as follows: (1) Dissolve 1 gm of hemoglobin (hemoglobin substrate, Worthington Co.) and 200 mg of tris (hydroxymethyl aminomethane base) in 10 ml of water, then filter through Whatman No. 42 filter paper. (Tris-buffered fresh red blood cell hemolysate containing 10 gm% hemoglobin instead of the hemoglobin solution can also be used.) (2) Cellulose nitrate solution is prepared as follows: collodion U.S.P. is evaporated to 20 percent of its orginal weight so that most of the ether and alcohol is removed; it is then made up to its original volume with ether (U.S.P. from R. Squibb & Sons, Ltd.). The presence of the original proportion of alcohol in the collodion solution is undesirable because it causes precipitation of the protein in the aqueous phase. (3) Organic solution: 100 ml of ether (U.S.P.) is saturated with water by shaking with distilled water in a separating funnel, then discarding the water layer. Then, 1 ml of Span® 85 (Atlas Powder Co.) is added.

To a 150 ml glass beaker containing 2.5 ml of tris-buffered hemoglobin solution is added 25 ml of the organic solution. The mixture is immediately stirred using a Jumbo magnetic stirrer (Fisher Co.) with a speed setting of 7 and a 4 cm stirring bar. After the emulsifier has been running for five seconds, 25 ml of the cellulose nitrate solution is added and stirring is continued for another one minute. The beaker is then covered and allowed to stand unstirred at 4°C for 45 minutes. During this period, the cellulose ester gradually pre-

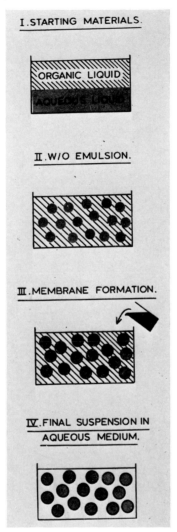

Figure 3. Summaries of typical procedures in the preparation of artificial cells.

cipitates at the interface of each microdroplet. If the microcapsules have sedimented completely by the end of the 45-minute period, all but 4 ml of the supernatant can be conveniently removed. (In some cases the suspension may have to be centrifuged at 350 g for five minutes before the supernatant can be removed.) After the removal of most of the supernatant, 30 ml of *n*-butyl benzoate (Eastman) containing 1% (v/v) Span 85 is added and stirred with the suspension on the jumbo stirrer (speed 5) for 30 seconds. The suspension is then allowed to stand uncovered and unstirred at 4°C for 30 minutes more to allow the ether to evaporate and the outer surfaces of the artificial cells to set.

To transfer the artificial cells from the organic-liquid phase into an aqueous phase, the supernatant should first be removed. This could be done readily if the artificial cells have sedimented completely, otherwise centrifugation is required. After the supernatant is removed, 25 ml of a dispersing solution prepared by dissolving 12.5 ml of Tween® 20 (Atlas Powder Co.) in an equal volume of water is then added to the suspension. The aqueous suspension is dispersed by stirring with the jumbo stirrer (speed 10 for 30 seconds). Stirring is slowed down (speed 5) and 25 ml of water added. The stirred suspension is further diluted with 200 ml of water. The slightly turbid supernatant may now be removed by sedimentation or centrifugation. The artificial cells are washed repeatedly in 1% Tween 20 solution until no further leakage of hemoglobin takes place. Repeated washings remove the more fragile artificial cells, leaving only those which are well formed (Fig. 4). These are suspended in a 0.9 gm% sodium chloride solution.

Variation In Contents

Proteins, enzymes, and other materials (in suspension or in solution) to be enclosed in the artificial cells are added to the 10 gm% hemoglobin solution. If the material is soluble, a filtrate (Whatman No. 42) of the final solution should be used. The remaining procedures are as described. Sufficient tris buffer should be added to the aqueous solution to maintain a pH of more than 8.5. All enzymes and proteins tested so far have been successfully microencapsulated when they are added to the hemoglobin solution. The hemoglobin, besides its necessity in the microencapsulation procedure, stabilizes the microencapsulated enzymes and proteins.

Variation In Diameters

Diameter of the artificial cells depends on the diameter of the aqueous microdroplets before membrane formation. Thus, the mean diameter of artificial cells in this procedure can be varied by the speed of the mechanical stirring (Fig. 5) or the presence or absence of the emulsifying agent Span 85. For pre-

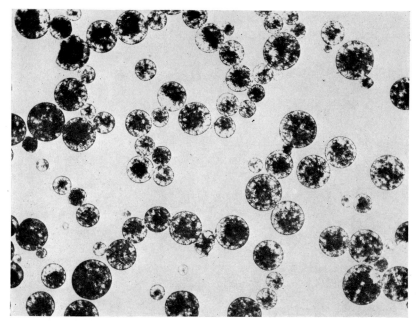

Figure 4. Collodion membrane artificial cells, mean diameter 80μ. (From Chang *et al.*, 1966. Courtesy of the National Research Council of Canada.)

paring artificial cells of diameter of less than 5μ (Fig. 6), more complicated procedures as described earlier (Chang *et al.*, 1966) are required.

Variation In Membrane Materials

Polymers other than cellulose nitrate can be used. However, modification of the procedures will be required. For instance, polystyrene membrane artificial cells can be formed using a modified procedure (Chang, *et al.*, 1966). Here, polystyrene is dissolved in benzene (0.2 gm/ml) and the procedures of organic phase separation carried out with benzene substituted throughout for ether, and polystyrene for cellulose nitrate.

INTERFACIAL POLYMERIZATION

Introduction

Morgan and his colleagues (1959, 1965) described their extensive studies in which they introduced interfacial polycon-

Artificial Cells

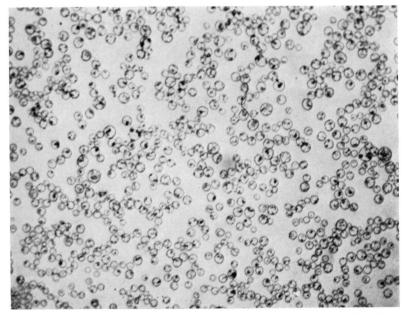

Figure 5. Collodion membrane artificial cells, mean diameter 5μ. (From Chang *et al.*, 1966. Courtesy of the National Research Council of Canada.)

densation to prepare nylon and other polymers at room temperature. Thus they demonstrated that a diamine (e.g. hexamethylenediamine) in the aqueous phase reacts at the interface with a diacid halide (e.g. sebacoyl chloride) in the organic phase to form nylon. Many other polymers can be prepared by interfacial polycondensation (Morgan, 1959, 1965); for example, polyamides, polyureas, polyurethanes, polysulfonamides, and polyphenyl esters.

We (Chang, 1964; Chang *et al.*, 1966) have adapted the principle of interfacial polymerization into the procedure (Chang, 1957, 1964) for the preparation of artificial cells (Fig. 3).

Typical Example

The starting materials are as follows. (1) Hemoglobin (hemoglobin substrate, Worthington Co.) solution (10 gm in 100 ml water) or fresh red blood cell hemolysate containing 10 gm% hemoglobin. (2) An "alkaline diamine solution" is prepared as a 100 ml aqueous

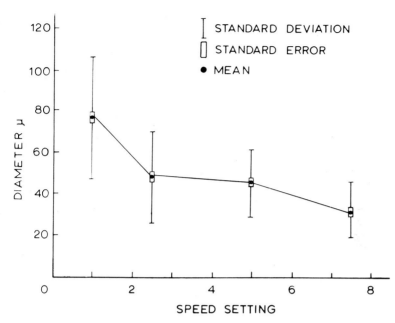

Figure 6. Collodion membrane artificial cells; effect of stirring speed on diameter. Span 85 concentration at 1% (v/v). (From Chang *et at.*, 1966. Courtesy of the National Research Council of Canada.)

solution containing 4.4 gm of 1,6-hexamethylenediamine (Eastman), 1.6 gm of $NaHCO_3$ and 6.6 gm of Na_2CO_3. This stock solution is filtered and then stored in the refrigerator at 4°C, but should be discarded after 24 hours or as soon as it becomes slightly cloudy. (3) A stock organic liquid to be made up of 1 part chloroform and 4 parts cyclohexane and containing 1% by volume of an emulsifying agent, Span 85, is prepared just before use. (4) "Sebacoly chloride organic liquid" 0.018M, is prepared within 20 seconds before use by adding 0.1 ml of pure sebacoyl chloride (Eastman) to 25 ml of "stock organic liquid." Failure to prepare satisfactory microcapsules is frequently due to the deterioration of sebacoyl chloride after the seal of the bottle has been opened. A new bottle of sebacoyl chloride will usually solve this problem. (5) Dispersing solution of Tween 20 is made up of equal volumes of water and Tween 20.

The hemoglobin or hemolysate solution (2.5 ml) is added to an equal volume of the alkaline diamine solution in a 150 ml beaker placed in an ice bath. The two solutions are gently mixed with a glass rod. Within ten seconds after mixing the two solutions, 25 ml of the stock organic liquid is added. The beaker with its ice bath is

immediately placed on a Jumbo magnetic stirrer, and a 4 cm stirring
bar is placed in the beaker. The speed is set at 5 and stirring is carried out for one minute. Then, while stirring is continued at the same
speed, 25 ml of the sebacoyl chloride organic liquid is added and
the stirring is continued. (The duration of the reaction determines the
characteristics of the membrane. If the reaction is too brief, a leaky
macroporous membrane is formed. On the other hand, prolonged
reaction yields a thick and less porous membrane.) For the standard
nylon membrane artificial cells, exactly three minues after the addition of the sebacoyl chloride organic liquid, the reaction is quenched
by the addition of 50 ml of the stock organic liquid. Stirring is dis-
continued, and the suspension is allowed to sediment, or it is centrifuged for 15 seconds at 350 g. All the supernatant is discarded,
50 ml of the 50% Tween 20 solution is added, and the suspension is
stirred at a magnetic stirrer speed setting of 10 for 30 seconds.
The speed is then decreased to 5, 50 ml of distilled water is added
to the stirred suspension, and stirring is continued for another 30
seconds. The suspension is then added to 200 ml of distilled water
and stirring is continued at the same speed for another 30 seconds.
200 ml of distilled water is then added, and stirring is continued at
the same speed for another 30 seconds. The suspension is allowed
to sediment or else to centrifuge for one minute at 350 g, and all the
supernatant is discarded. Alternatively, the suspension may be sieved
in a mesh screen to remove the supernatant. The artificial cells are
finally resuspended in 0.9 gm% NaCl. The artificial cells may be
crenated in the hypertonic Tween 20 solution; but they regain
sphericity (Fig. 7) in the 0.9% NaCl if the period of contact with
the strong Tween 20 medium has been kept as short as possible.

Some comments should be made about the choice of solvents
and reactant concentrations. Although many organic liquids may
be used in interfacial polymerization, the combination of chloroform-cyclohexane which we have used satisfies the following
requirements. First, it should not readily denature proteins
when used in the given proportion. Secondly, it is desirable to
have a liquid with specific gravity fairly close to 1. Thirdly, the
partition coefficient of the organic liquid for the diamine determines the properties of the membrane (Morgan, 1965). Chloroform, which has a high affinity for diamine, produces strong but
coarse and thick membranes, whereas cyclohexane which has a
low affinity for diamine produces smooth and very thin but weak
membranes. We found that these two solvents combined in the

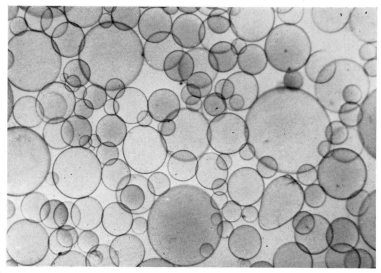

Figure 7. Nylon membrane artificial cells, mean diameter 90μ. (From Chang, 1967. Courtesy of *Science Journal*, London.)

stated proportion give a fluid which has a satisfactory specific gravity (0.91) and allows the production of a thin but sufficiently strong membrane. Finally, carbon tetrachloride and benzene, though capable of forming strong and useful membranes, are not used because of their possible toxic effects.

Some comment should be made about the optimal ratio of the two reactants in the polymerization, hexamethylene-diamine and sebacoyl chloride. The significant ratio here is that of the concentrations of the reactants in their respective solvents, rather than that of their total quantity in the system as whole. Our selection of an optimal concentration ratio was partly based on studies of Morgan (1959), who determined the ratio that produces polymers of the highest molecular weight. Morgan's results were obtained mostly with an unstirred interface, where the aqueous phase contained only diamine with NaOH as the hydrogen ion acceptor. In our case, however, the aqueous phase contained proteins like hemoglobin, and enzymes and bicarbonate-carbonate buffer, and the reaction was carried out at the interface of the aqueous microdroplets. To prepare the best membrane, the optimal concentration ratio in our special case was a diamine to acid-chloride ratio of 22:1. It should be emphasized that any variety in the reaction condition or in the composition of the material to be enclosed may alter the optimal condition, and this will require readjustment.

It should be noted that the presence of a 10 gm% hemoglobin solution greatly strengthen the membranes of the artificial cells. A small fraction of the hemoglobin at this concentration participates in the interfacial polymerization. This results in cross-linking of the linear nylon polymer, thus greatly strengthening the membrane. Similar observations have later been made by Sparks *et al.* (1969) and Shiba *et al.* (1970). In addition, the high concentration of protein also stablizes some enzymes.

Control Of Diameters

Droplet size in mechanically prepared emulsion is never uni-

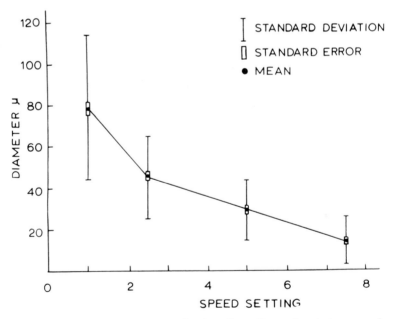

Figure 8. Nylon membrane artificial cells; effect of stirring speed on diameter. Span 85 concentration at 15% (v/v). (From Chang *et al.*, 1966. Courtesy of the National Research Council of Canada.)

form. Thus the artificial cells formed have a diameter with a size distribution of ± 50% SD. However, using sieves (Fisher Co.) of different meshes, artificial cells with a very narrow size distribution (SD ± 5%) have been obtained (Chang and Poznansky, 1968c).

The mean diameter of artificial cells produced by this procedure can be varied by changes in the degree of mechanical emulsion, the amount of chemical emulsifier, and the viscosity of the continuous phase. This is illustrated in Figures 8 and 9.

Very small artificial cells of less than 5μ diameters can be prepared by using a more powerful emulsifier, e.g. Virtis homogenizer. However, in these cases more complicated procedures as described in Chang, 1964 and Chang *et al.*, 1966 have to be followed.

Very large artificial cells may be obtained by emulsifying in a high-viscosity continuous phase (Chang, 1965), e.g. one of the

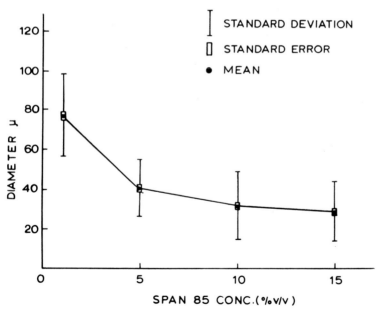

Figure 9. Nylon membrane artificial cells; effect on diameter of varying Span 85 concentration. (From Chang *et al.*, 1966. Courtesy of the National Research Council of Canada.)

silicone fluids, which are available in a wide range of viscosities. Since sebacoyl chloride is insoluble in these silicone fluids, it is added to a small volume of the stock organic solution (0.18M). This is then added to 10 volumes of silicone oil. In this way, artificial cells of up to 0.5 cm diameter may be obtained with silicone fluid (Dow Corning 200 fluid) with the Jumbo magnetic stirrer set at 1, and with no Span 85 in the stock organic liquid. Large artficial cells can also be formed by the drop technique.

Drop Technique

Large artificial cells of uniform diameters can be prepared by a very simple procedure (Chang, 1965; Chang *et al.*, 1966). Sebacoyl chloride organic liquid (0.018M) is prepared as before, but Span 85 is omitted from the organic solution. Alkaline diamine solution with or without hemolysate solution or hemoglobin solution is prepared as described before. The aqueous solution is added dropwise through a needle placed about 1 cm above the surface of the sebacoyl chloride solution. This can be done manually, or by using a Harvard Perfusion Apparatus for continuous drop formation. After a reaction time of five minutes, during which the solution is shaken gently, the membrane-enclosed droplets can be resuspended into an aqueous medium by a number of procedures, depending on the nature of the enclosed material. Supernatant of organic solution can be decanted, and any trace amount allowed to evaporate completely before adding saline to suspend the droplets. Another way is to remove most of the organic solution, then add liquid which is soluble both in the organic and the aqueous phase (e.g. alcohol, acetone, dioxane) to wash out the organic solvent and to allow resuspension into saline (Chang, 1965). Another way is to use a Tween 20 solution for this transfer. Material to be enclosed is added to the alkaline diamine solution before the liquid is dropped into the sebacoyl chloride solution, for instance, hemolysate solution, protein, enzymes, ion exchange resin.

An electrical emulsifier like the one described by Nawab and Mason (1958) has also been used to form microdroplets of uniform diameter for the drop technique (Chang, 1965). More recently the drop technique has been extended and modified by

Sparks' group (Sparks *et al.*, 1969). Here the aqueous phase (containing 1,6-hexanediamine (0.2M), NaOH (0.4M) and hemolysate) is forced through a capillary or hypodermic needle. A gas at constant pressure is fed through a larger concentric capillary, moving past the inner capillary at constant velocity. As a droplet begins to form, its surface extends into the annular stream of gas and is detached by the drag force of the stream of gas. The size of the droplet is controlled by the velocity of the annular gas stream. The droplets fall into a tube (60 cm in length, 1 cm in diameter) containing cyclohexane, carbon tetrachloride, Span 85, and sebacoyl chloride. A gradient of sebacoyl chloride exists in the tube, zero concentration at the top to assist efficient droplet entry, and maximal sebacoyl chloride concentration (0.045M) in the collection flask at the bottom of the tube. Artificial cells with mean diameter between 100μ and 1000μ have been formed this way; size distribution is in the order of $\pm$ 14%.

Variation In Membrane Materials

The procedure described for nylon membrane artificial cells gives the basic principle of preparation. Other diamines besides hexamethylenediamine and other diacids besides sebacoyl chloride can be adapted into this procedure (Chang, 1964; Chang *et al.*, 1966). This includes cross-linked protein membranes (Chang, 1964; Chang *et al.*, 1966), membranes with sulfonic acid groups (Chang, 1964; Chang *et al.*, 1966; Koishi *et al.*, 1969), polyurethane and polyphenolester membranes (Suzuki *et al.*, 1968), and polyphthalamide membranes (Koishi *et al.*, 1968).

Variation In Content

As in the case of collodion artificial cells, material to be enclosed can be added to the diamine-hemolysate mixture before enclosure (Chang, 1964, 1965; Chang *et al.*, 1966, Sparks *et al.*, 1969). However, unlike the case of collodion artificial cells, a number of enzymes become inactivated by the presence of hexamethylenediamine. For example, uricase and catalase lose most of their activity after enclosure in nylon artificial cells, although they remain active after enclosure in collodion artificial cells. Other enzymes like urease and asparaginase retained good activity after enclosure. In the case of urease, 200 mg of urease (Nutritional Biochemicals Corp., soluble urease, 1.5 Sumner units/mg) can be dissolved in 3 ml of the hemolysate-diamine mixture, the rest of the procedure being as described. Proteins, macromolecules, or suspensions can be added to the hemolysate-diamine mixture before enclosure. In the drop technique, the artificial cells can be prepared to contain only NaOH, and other materials without the presence of hemoglobin.

SECONDARY EMULSION

Principle

Secondary emulsion is a well-known phenomenon in physical chemistry. For example, a fine emulsion of aqueous microdroplets is dispersed in an organic phase. The organic phase containing the fine aqueous microdroplets is then dispersed in the form of larger droplets in an aqueous phase. The final product is thus an aqueous suspension of organic phase droplets, each of the droplets containing finer aqueous microdroplets. If the organic phase contains a polymer-forming solution it can be solidified to form solid spheres, each containing a number of aqueous microdroplets. Unlike those described in the earlier section, this procedure does not give spherical ultrathin membranes enclosing an aqueous interior. Instead it results in solid polymer micorspheres containing from zero to several aqueous microdroplets. The requirements for double emulsion are such that a large proportion of the solid microsphere consists of the polymer materials; also, the aqueous content varies greatly from one microsphere to another. Examples of artificial cells formed from this procedure are given below.

Silicone Rubber

The silicones are available in the form of fluids, resins, and rubbers. A thick liquid polymer, Silastic RTV, (Dow Corning), can be vulcanized at room temperature by the addition of a catalyst (dibutyl tin dilaurate or stannous octoate) to form a rubber that has useful mechanical properties and is chemically and biologically inert (Braley, 1960). The properties of silicone rubber do not permit its use as a dialyzing membrane because the material, though it can be formed into membranes, is extremely impermeable to water and aqueous solutes. However, its affinity for oxygen and carbon dioxide is extremely high (150 to 300 times more than polyethylene and 800 to 1000 times more than cellulose acetate). It is also permeable to some lipid-soluble materials. Thus, Folkman and Long (1963) found that lipid-soluble drugs (e.g. triiodothyronine and digitoxin) sealed in closed Silastic tubes of 3.5 mm internal diameter could penetrate the Silastic wall by solvation.

These desirable properties led to the preparation of artificial cells with silicone rubber (Chang, 1966). Silicone rubber (Silastic) artificial cells can be readily prepared as follows: protein solution, suspension, powder, or other materials added dropwise is emulsified in 2 volumes of Silastic (Dow Corning S-5392); while stirring is continued, 0.025 volumes of stannous octoate catalyst (Dow Corning) is added. This is followed immediately (within two seconds) by 20 volumes of 50% Tween 20 solution containing 50 mg/100 ml of sodium hydrosulfite. Stirring is then continued at a slower speed for 20 minutes or more to attain the "tack free time" of the Silastic polymer. Hemoglobin in Silastic rubber microspheres prepared without too vigorous stirring remains unaltered and can combine reversibly with oxygen, and it retains this ability for at least a few days of cold storage. Other liquids or particles can also be similarly entrapped within silicone rubber microspheres.

Other Polymers

The principle of secondary emulsion has also been used to prepare microspheres with cellulose polymer material containing erythrocyte hemolysate (Chang, 1965). This principle has also been used to form artificial cells by another group of workers in Japan (A. Kondo, 1968; Kitajima *et al.*, 1969). Like our procedure of secondary emulsion, an aqueous solution of hemoglobin or enzymes is emulsified in a solution of a film-forming polymeric material (e.g. polystyrene, ethyl cellulose, dextran stearate dissolved in organic solvent like benzene or ethyl ether), forming a water-in-oil emulsion. This emulsion is then emulsified in an aqueous solution consisting of 1%-4% gelatin solution and 0.5%-0.02% anionic or nonionic surfactant solution to form the secondary emulsion. Finally, the temperature of the emulsion is raised up to 40°C for one to two hours to evaporate the organic solvent.

MEMBRANES WITH FIXED CHARGE

Collodion membrane artificial cells prepared as described (Chang, 1957, 1964, 1965) contained a negative charge because of the carboxyl groups in the polymer. Nylon membrane artificial cells (Chang *et al.*, 1963, 1966; Chang, 1964) contain both carboxyl groups and amino groups. By the proper selection of polymer materials, membranes with the desired fixed charge can be obtained. For example, artificial cell membranes with strong negative charge groups have been prepared by using a sulfonated diamine such as 4,4'-diamine-2,2'-diphenyldisulfonic acid (Chang, 1964, 1965; Chang *et al.*, 1966). Thus "sulfonated-nylon" membrane artificial cells of varying surface negative charge can be prepared by adding different amounts of the sulfonated diamine (e.g. 3mM/liter) to the alkaline 1,6-hexanendi-

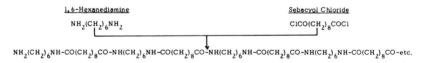

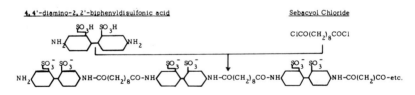

Figure 10. *Upper,* schematic representation of nylon artificial cell membrane. *Lower,* schematic representation of sulfonic nylon artificial cell membrane.

amine solution, then following the standard procedure described. The reaction of hexamethylenediamine and 4,4′-diamino-2,2′-diphenyldisulfonic acid with sebacoyl chloride may be represented by the equation in Figure 10. Koishi *et al.* (1969) supported and extended our approach for the preparation of sulfonated membrane artificial cells by using a combination of diethylene-diamine and 4,4′-diaminostilbene-2,2′-disulfonic acid in varying ratios. A number of ion-selective membranes (Chang, 1965) can also be used to incorporate into the membrane of the artificial cell system. Polysaccharides with strong sulfonated groups have also been successfully incorporated into the membrane (Chang *et al.*, 1967b). Quaternarisation or addition of cationic group to the membrane gives strong positively charged membranes.

MEMBRANES OF PROTEIN, LIPID, AND POLYSACCHARIDE

Introduction

In nature, cell membrane is made up of a complex lipoprotein structure coated externally by mucopolysaccharide. The lipoprotein portion is classically considered to be arranged in a bimolecular pattern, with two monolayers of lipid molecules sandwiched between monolayers of nonlipid molecules (Davson and Danielli, 1943; Robertson, 1960). Alternative theoretical models of cell membranes include a unit membrane with the lipid component in the form of globular micelles surrounded by the protein component of the membrane (Sjostrand, 1963; Lucy, 1964) or a protein matrix coated with lipid (Lenard and Singer, 1966; Korn, 1968). It is clear that at present, the exact structure of the lipoprotein complex is far from being established.

As a further step towards reconstitution of biological cells besides use of the contents of biological cells, the artificial cell membranes have also been formed from biological materials. Thus, cross-linked protein (Chang, 1964, 1965; Chang *et al.*, 1966), absorbed protein (Chang, 1969e), lipid-polymer or lipid-protein complex (Chang, 1969b), lipids (Mueller, 1968; Pagano and Thompson, 1968), and polysaccharides (Chang *et al.*, 1967b) have all been incorporated into the artificial cell membanes.

Artificial Cell Membranes Of Proteins

Protein is an important component of biological membranes. Thus in the preparation of artificial cells, protein has been incorporated into the membranes. Since proteins are in fact polyamines containing free amino groups, protein molecules have been incorporated into the artificial cell membranes by cross-linking with diacids. Thus, in the standard procedure of preparing nylon membrane artificial cells (Chang, 1964, 1965; Chang *et al.*, 1966; Shiba *et al.*, 1970) the membrane contains cross-linked protein (Fig. 11). The presence of protein in these membranes has been found to greatly strengthen the artificial cell membranes, since it serves to cross-link the linear nylon polymer molecules. Artificial cell membranes have also been formed entirely of cross-linked protein (Chang, 1964, 1965; Chang *et al.*, 1966). Here, no alkaline diamine is used, and only 10 gm% hemoglobin solution is used. If the standard procedure is to be used, very small artificial cells of less than 5μ, as described in detail elsewhere (Chang *et al.*, 1966), have to be formed in order for the cross-linked protein membranes to have sufficient mechanical strength. If larger artificial cells are to be formed, the drop technique should be used. Here again, the alkaline diamine solution is omitted and only the 10 gm% hemoglobin solution is used. In this case, the concentration of the sebacoyl chloride in the sebacoyl chloride organic liquid is increased to 0.5M from the original 0.018M.

Proteins can also be incorporated into artificial cell membranes by physical absorption. Thus, albumin has been incorporated into artificial cell membranes by physical adsorption (Chang, 1969e; Chang *et al.*, 1970), making the artificial cells biologically compatible. The details of this procedure will be described in a later chapter. Physical adsorption can be combined with glutaraldehyde treatment to give a more permanent membrane. Another way that may be used is to covalently bind the protein.

Na^+,K^+-activated SATPase, an important enzyme system in biological cell membranes has also been incorporated into the artificial membrane (Chang and Rosenthal, 1969). Here Na^+,K^+-activated SATPase is prepared from red blood cell membranes. Artificial cells are suspended in the Na^+,K^+-activated SATPase.

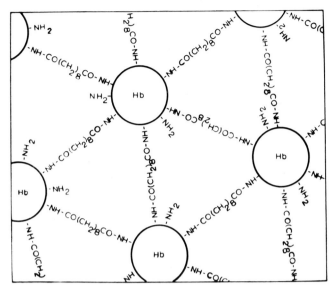

Figure 11. Schematic representation of cross-linked protein artificial cell membrane. Hb—protein molecules.

The amount of SATPase adsorbed onto the artificial cell membrane is increased further by the removal of calcium in the SATPase solution. This way, more SATPase is precipitated on the artificial cell surface. A significant amount of SATPase activity is present in the membranes of artificial cells prepared this way.

Artificial Cell Membranes With Polysaccharides

Mucopolysaccharides coating biological cells are important components of cell membranes (Cook *et al.*, 1961; Rambourg *et al.*, 1966). Our interest in biologically compatible membranes has led us to prepare artificial cell membranes with a polysaccharide, heparin (Chang *et al.*, 1967b, 1968). This has been prepared by a number of procedures, the details of which will be discussed later under the appropriate chapter.

Artificial Cell Membranes Of Lipids

Mueller and Rudin (1968) used a modification of our standard procedure (Chang, 1964) to prepare artificial cells with lipid membranes enveloping erythrocyte hemolysate or other proteins. These are prepared as follows (Müeller and Rudin, 1968): 0.1 volume of 1% to 10% erythrocyte hemolysate is added to 0.1% to 1% solution of brain or beef heart lipids in hexane or octane. The mixture is emulsified by shaking, then centrifuged at 300 to 500 rpm. Most of the organic solvent is removed, then 10 volumes of 0.1M NaCl is added to the lipid-enclosed artificial cells. This is gently shaken, centrifuged, and washed three times to separate the artificial cells from the lipid layer and to remove the hexane. Aggregation or separation of these artificial cells can be controlled by calcium and EDTA. Other organic solvents like triolein or alcohol can be used. Other solutions like egg white or cytoplasmic constituents can be enclosed. Electromicroscopic pictures show that the enclosing lipid membranes are usually 60 to 100 Å thick with a typical osmic stain. Because of their lipid composition and their ultrathin membranes, artificial cells formed this way are rather fragile, and remain intact for only a short period of time (Mueller and Rudin, 1968).

Pagano and Thompson (1967, 1968) prepared large spherical ultrathin lipid membranes (a few millimeters in diameter) using the drop technique. Unlike the artficial cells reported in this monograph, these spherical ultrathin membranes contain only sodium chloride solution. First a NaCl density gradient (1.0 to 2.0 gm/ml) is formed and kept at 36°C ± 1 degree. A microburette syringe connected to a polyethylene tube is filled with a 3.5M NaCl solution. A lipid solution (containing 4 gms egg phosphatidyl choline, 28 ml n-tetradecane, 42 ml chloroform and 28 ml methanol) is used to coat the end of the polyethylene tubing. The tubes are placed into the NaCl density gradient. A droplet of the NaCl solution is discharged from the tubing. The droplet, which is coated with the lipid solution at the end of the polyethylene tubing, falls through the density gradient, resting at a given point. Spherical membranes of about 4 mm are formed this way. Initially the lipid membrane is more than 2000 Å in thickness. However, the lipid phase gradually moves to the top of the sphere to form a cap, leaving the remaining sphere a very thin lipid membrane. Using light reflections, the membrane thickness was found to be about 90 Å.

Artificial Cell Membranes Of Lipid—Macromolecule Complexes

Biological cell membranes are not made up solely of protein or solely of lipid; instead, they are made up of a lipoprotein complex. Thus artificial cell membranes formed of a lipoprotein or even a lipid-polymer complex will be more similar to biological cell membranes. Studies have been carried out along these lines. Thus nylon membrane artificial cells or cross-linked protein membrane artificial cells have been prepared and then complexed with naturally occurring lipids (Fig. 12) (Chang, 1969b). The artificial cells are first formed by the standard drop technique described earlier but omitting the final steps of transfer into the aqueous phase. Instead they are left in the organic liquid. The organic liquid is removed and the artificial cells washed with *n*-tetradecane, then suspended in a *n*-tetradecane solution containing an equimolar amount of cholesterol-lecithin.

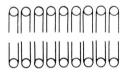

A

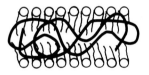

B

A - Lipid molecule

CX - Polymer or protein

Figure 12. *Upper,* proposed arrangement of lipid molecules in spherical ultrathin lipid membranes; similar to the classical bimolecular lipid theory for biological membranes. *Lower,* proposed arrangement of lipid-protein or lipid-polymer artificial cell membranes.

This is left for one hour. Then, most of the supernatant is removed. The cells are than gently transferred into a Ringer solution. By gently stirring, excess lipid solution floats to the top, whereas the artificial cells with lipid-macromolecule complex membranes settle to the bottom. Further thinning of the lipid takes place as excess lipid gradually accumulates at the top of the artificial cells. Macrocyclic molecules, like Valinomycin, can be incorporated into the lipid component. They are either dissolved in the lipid solution before membrane formation, or added to the aqueous suspension. Further details will be described in papers being prepared.

MICROENCAPSULATION OF PARTICULATE MATTERS

Principle

Particulate matter may be enclosed in artificial cells. One way is to add the suspension to the aqueous phase at the outset of the procedure (Chang, 1964, 1965, 1966; Chang *et al.*, 1966, 1967b). Thus, no particular difficulty was encountered in this laboratory with the encapsulation of suspensions like cell hemogenates, microcrystalline materials, activated charcoal, resin, insolubilized enzymes, and others. Another way to enclose biologically active particulate matter is by polymer coating (Chang, 1969e; Chang *et al.*, 1968, 1970). A few typical examples are discussed in the following sections.

Microcapsules Within Artificial Cells

Nylon membrane artificial cells of 20μ mean diameter containing hemolysate were prepared by the usual procedure (stirring speed of 5 and Span 85 concentration of 15% v/v). A 25% (v/v) suspension of these artificial cells was placed in the hemoglobin-diamine-buffer mixture prepared as described and then encapsulated by the standard procedure, with a stirring speed of 1 and Span 85 concentration of 1% (v/v). Such artificial cells are shown in Figure 13. Mean diameter was about 20μ for the enclosed microcapsules and about 100μ for the artificial cells. The enclosed microcapsules might be considered to be analogous to intracellular organelles. If the outer artificial cells are ruptured by pressing on the overlying cover-slide, the enclosed micro-

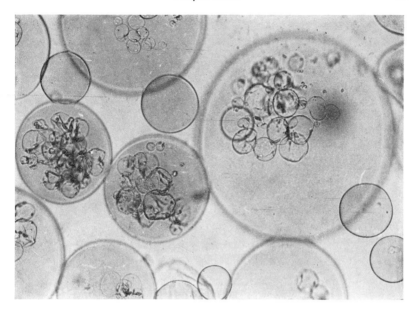

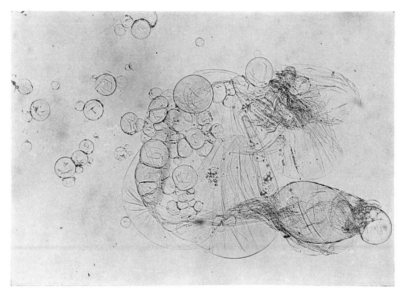

Figure 13. *Upper,* nylon membrane artificial cells enclosed in larger ones. (From Chang *et al.,* 1966. Courtesy of the National Research Council of Canada.) *Lower,* outer membrane ruptured mechanically, releasing enclosed artificial cells.

capsules can be observed under the microscope to flow out through the ruptured outer artificial cells.

Microencapsulation Of Intact Cells

The microencapsulation of mammalian erythrocytes using the interfacial polymerization technique is a typical example. The standard technique for making large nylon membrane artificial cells cannot be used for enclosing erythrocytes for two reasons: (1) erythrocytes undergo at least partial lysis when emulsified in the stock organic liquid of chloroform and cyclohexane, and (2) erythrocytes are rapidly and completely lysed by the alkaline diamine solution. After a number of trials, the following solution was found for these problems (Chang, 1965). The erythrocytes were suspended in an isotonic solution of hemolysate containing 10 gm% hemoglobin rather than in the diamine solution, and a silicone oil (Dow Corning 200 Fluid) was substituted for the stock organic liquid. The interfacial polymerization was then carried out using silicone fluid as described earlier for cross-linked protein membranes.

Figure 14 shows a large number of human erythrocytes suspended in hemolysate within an artificial cell of about 500μ diameter prepared by the drop technique.

Artificial Cells Containing Synthetic Particulate Material

Biologically active synthetic particulate matter can be enclosed in artificial cells (Chang, 1966, 1969e; Chang *et al.*, 1967, 1968, 1970; Levine and LaCourse, 1967; Sparks *et al.*, 1969). Examples are the enclosure of ion exchange resins and activated charcoal. The exact details of the procedure will be described in the chapter about detoxicants.

INDUSTRIAL MICROENCAPSULATION TECHNOLOGY

There is a substantial amount of very elegant industrial technology in microencapsulation which until recently appeared mostly in patent literature. Industrial microcapsules are prepared with special properties to suit their particular applications and as a result are different from the type of artificial cells described in this monograph. For instance, the industrial microcapsules

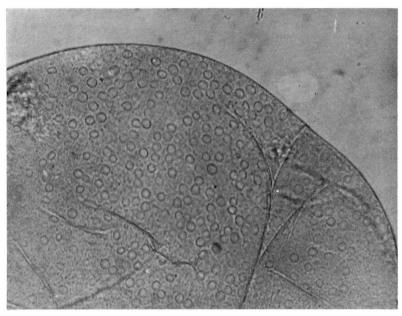

Figure 14. Human erythrocytes suspended in hemolysate and enclosed in a large artificial cell with cross-linked protein membranes. (From Chang *et al.*, 1966. Courtesy of the National Research Council of Canada.)

function as microscopic inert containers, separating the enclosed materials completely from the external environment. The enclosed material can act only when the enclosing wall is disrupted, releasing the enclosed material. The best known example is the NCR carbonless carbon paper. Here, microcapsules containing oil are coated on paper. Mechanical pressure ruptures these microcapsules, releasing their oily contents to interact and form a print. Other examples are the microencapsulation of pharmaceuticals, nuclear fuel particles, battery separators, food products, perfumes, adhesives, agricultural products, petroleum, and others. In all cases, the enclosing walls of the industrial microcapsules are made as impermeable as possible. They differ from the artificial cells described in this monograph in a number of important points. In artificial cells, the enclosed materials are biologically active materials like enzymes, other proteins, and detoxicants. The enclosed materials of artificial cells do not depend on being released from the enclosing membrane for action. While remaining at all times enveloped by the enclosing membrane and prevented from coming into direct contact with the external environment, the enclosed materials act on external permeant molecules diffusing into the artificial cells. The enclosing membrane of artificial cells is prepared in such a way that while impermeable to macromolecules or suspensions, it is extremely permeable to most of the solutes normally present in the biological fluid. This high permeability is a result of the ultrathinness (200 Å) and porosity (radius of 16 Å) of the enclosing membrane of the artificial cells. Despite these major differences between industrial microcapsules and the artificial cells described in this monograph, one should not overlook the tremendous potentiality of using the available technology of industrial microencapsulation for preparing artificial cells. This technology has not previously been used in the preparation of the type of artificial cells described in this monograph. After the clinical potentiality of artificial cells has been demonstrated by studies in our laboratory, the possibility of applying the industrial microencapsulation technology for the large-scale production, improvement, or modification of artificial cells is beginning to be explored. Thus, a very brief introduction to industrial microencapsulation technology might be useful. This monograph covers only published scientific literatures and not the patent literatures. The various patented industrial microencapsulation procedures can be found in recent reviews (*Chemical Week*, 1965; Flinn and Nack, 1967; Nack, 1970; Luzzi, 1970; Herbig, 1968; Bakan and Anderson, 1970). Very briefly, the typical groups are as follows. First, there is the aqueous phase separation principle developed by Green and Schneidcher of the National Cash Register Company (1957). This is the first industrial microencapsulation technology developed. This technique is limited to the encapsulation of water-immiscible liquids or solids. A typical example is to form emulsion of an oil in an equeous medium containing dissolved gelatin. By a change in the composition, pH, or temperature, the gelatin is deposited on the emulsified oil droplets and then hardened. The best known practical example is the NCR carbonless carbon paper. Secondly, there is the organic phase separation principle involving the precipitation of dissolved polymer membranes on emulsified aqueous microdroplets; this is limited to the encapsulation of aqueous solutions or solid particles. The same principle has been used in this laboratory since 1956 (Chang, 1957) for the preparation of artificial cells with collodion membranes. In industry this principle has been used for microencapsulation by the procedure of nonsolvent addition (IBM in 1964) and polymer incompatibility (NCR in 1964). Thirdly, there is the interfacial reaction process similar to that first used in this laboratory for the preparation of nylon membrane artificial cells (Chang *et al.*, 1963, 1964, 1966). In industry it is used for the coating of fibers by reacting adsorbed catalyst with ethylene gas bubbles, encapsulating aerosol particles by *in situ* polymerization of monomers (Standard Research Institute in 1964), and encapsulating various materials with a dense impervious coating by chemical vapor deposition. The IIT Research Institute (in 1964) produced aerosol microcapsules by the principal of electrostatic encapsulation, which means that one type of aerosol is coated by another with opposite charge and than allowed to solidify. Finally, there are numerous physical methods; a few of these might be mentioned. A large number of spray-coating methods are being used. There is the spray-drying technique where the material to be microencapsulated is suspended in the vaporizable solvent in which the enclosing material is dissolved. By spraying this into a heated chamber, the solvent evaporates, leaving a film on the particles. This has been used since 1962 for colour-forming systems for the office copying field. The fluid-bed spray-coating encapsulation used at Battelle-Columbus involves fluidizing core particles by a stream of air or other gas, spraying the particles with the wall material, and evaporating the solvent. Liquid can be frozen and the particles spray-coated. In the diffusion exchange microencapsulation method, microcapsules with solid cores are formed by spray; the cores are then exchanged for a fluid by diffusion, followed by coating of the microcapsules with a material which is impermeable to the fluid. Particles with a temperature above the melting point of the coating material cause the material to coat the particles by fused coating. In the extrusion method, two streams consisting of the coating material and the core material are extruded either by centrifugal force or by an extrusion nozzle device, resulting in the coating of the core material. Other physical methods are vacuum metalizing processes and capsulet processes. The exact details of nearly all industrial microencapsulation technology have not been published, but they can be found in patent literatures.

Chapter 3

BIOPHYSICAL PROPERTIES

MECHANICAL PROPERTIES

A RTIFICIAL cells prepared by the methods described and
suspended in an aqueous medium are usually spherical.
Preliminary observations suggest the membrane thickness in 80μ
nylon membrane artificial cells made by the standard procedure
is about 200 Å. That this ultrathin membrane is extremely flexible is demonstrated when the artificial cells are ruptured by dis-
secting needles to remove the contents. The membrane left is
extremely flexible and does not retain the original spherical
shape.

The sphericity of the artificial cell membrane is due to the
colloid osmostic pressure gradient generated by the enclosed protein. Thus, when spherical artificial cells are placed in a hypertonic solution, the outward movement of water results in the
folding of the spherical membranes (Fig. 15). In addition, when
spherical nylon membrane artificial cells of about 100μ diameter
are forced under hydrostatic pressure to flow slowly through the
narrow portion of a glass tube, the membrane is seen to fold
(Fig. 16). After passing through the narrow portion of the tubing, the artificial cells gradually return to their spherical shape.
The time course of the change makes it apparent that water is
forced out of the artificial cells as they pass through the con-
striction and returns after the artificial cells have emerged.
Another observation illustrating the flexibility of the artificial
cell membrane is shown in Figure 17. Here two large nylon
membrane artificial cells (diameter about 1 mm) are suspended
in silicone fluid and subjected to sheer stress and collision in a
Couette flow situation. In Figure 17 it can be seen that artificial
cells which are under sheer stress applied in opposite directions
elongate in the line of sheer stress; and on collision, the membranes flatten at the point of contact.

Jay and Edwards (1968) have used a cell elastometer to study

43

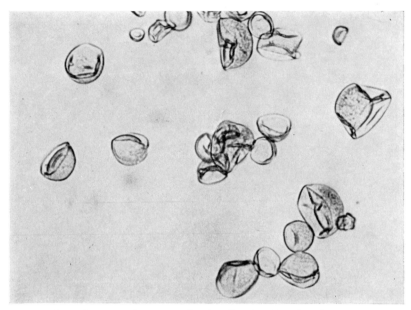

Figure 15. Nylon membrane artificial cells shortly after being placed in hypertonic solution. (From Chang, 1964. Copyright 1964 by the American Association for the Advancement of Science.)

more quantitatively the mechanical properties of nylon membrane artificial cells prepared by the standard procedure (Chang, 1964; Chang *et al.*, 1966). Their cell elastometer is similar to that used for studying the mechanical properties of urchin egg membranes (Mitchison and Swann, 1954) and erythrocyte membranes (Rand and Burton, 1963). The apparatus (Fig. 18) consists essentially of a micropipette (radius, R_p of 50μ to 260μ) connected to a mercury column which can be varied to adjust the pressure (P_1) inside the micropipette. Microcapsules of radius R_c are suspended in an aqueous medium having a hydrostatic pressure of P_2. The micropipette is then applied to the surface of an artificial cell, and by varying the negative pressure in the micropipette, a small tongue of membrane of length x can be drawn into the micropipette (Figs. 19 and 20).

A small negative pressure of 1 mm Hg can easily draw the whole artificial cell into the micropipette, if the artificial cell membranes are not under the tension of the osmotic pressure gradient. Thus the bending stress of the artificial cell membrane is negligible. On the other hand, when artificial cell membranes

Figure 16. Deformability of a nylon membrane artificial cell. The artificial cell, about 100μ in diameter and suspended in saline, is moving from left to right along a tapering glass capillary and is subjected to a hydrostatic pressure gradient from left to right. Note flattening of upstream surface, bulging of downstream surface, longitudinal folding of membrane, and smaller volume of artificial cell as a result of filtration of fluid through downstream surface into capillary. (From Chang, 1965; Chang *et al.*, 1966. Courtesy of the National Research Council of Canada.)

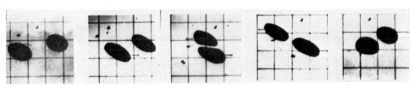

Figure 17. Deformability of nylon membrane artificial cells. Two artificial cells about 1 mm in diameter, suspended in silicone oil and placed in Couette flow apparatus. *Left to right,* zero flow; onset of Couette flow; collision; separation after collision; cessation of flow. (From Chang, 1965.)

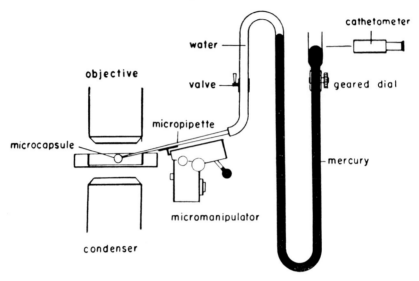

Figure 18. Schematic representation of cell elastometer. (From Jay and Edwards, 1968. Courtesy of the National Research Council of Canada.)

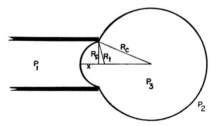

Figure 19. Schematic representation of artificial cell with a portion of the membrane drawn into the elastometer micropipette. Parameters involved are described in detail in text. (From Jay and Edwards, 1968. Courtesy of the National Research Council of Canada.)

are under an osmotic pressure gradient, the hydrostatic pressure required is as follows. When the pressure difference P_2-P_1 is plotted against the length of the tongue of the microcapsule membrane (x) sucked into the micropipette, the characteristic curve is shown in Figure 21. Initially, the relationship is linear, but with a further increase in the pressure difference (P_2-P_1), a point is reached when the slope of the curve suddenly decreases. Further increase in pressure beyond this point results in ruptur-

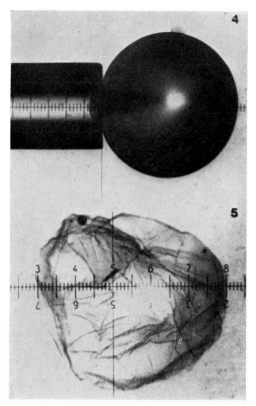

Figure 20. *Upper,* microphotograph of artificial cell with a portion of the membrane drawn into the elastometer micropipette (a much higher magnification was used in the experiments). *Lower,* a ruptured artificial cell showing the absence of any rigidity in the membrane. Spherical shape is maintained by the colloid osmotic pressure. (From Jay and Edwards, 1968. Courtesy of the National Research Council of Canada.)

ing of the artificial cells. Using equation $P_2-P_1=2T$, the membrane tension (T) at which the slope of the curve suddenly decreases is 2520 ± 20 dynes/cm. This same value is obtained for artificial cells of different diameters. The y-intercepts of these pressure deformation curves correspond to the pressure differences across the artificial cell membrane when the artificial cells are not deformed by the micropipettes. This is the internal pressure of the artificial cells (P_3). Micropipettes of different radii (R_p) tested on artificial cells of the same radius (R_c) give the same y-intercepts, thus the same internal pressures. On the other hand, when micropipettes of the same radius are used on artificial cells of different radii, different values of internal pressures (y-intercepts) are obtained (Fig. 21). When internal pressures are plotted against the reciprocals of the artificial cell radii, a straight line relationship is obtained, showing that the internal pressure is in inverse relationship to the diameter. From the equation $P_3-P_2=1T_c/R_c$, a membrane tension of 1840 dynes/cm is obtained. It should be noted that in these tests the high membrane tension is due to the osmotic pressure gradient across the membrane. When this osmotic pressure gradient is removed, e.g. in crenated artificial cells, this tension is no longer present.

In summary, Jay and Edwards show that the membranes of artificial cells prepared from different batches having a range of diameters all appear to have the same stiffness, the same tension, and the same critical tension—a remarkable uniformity in the mechanical properties of artificial cells. Despite the ultrathin membrane, it can withstand a membrane tension of up to 2520 ± 20 dynes/cm.

OPTICAL PROPERTIES

Anderson and Selkey (1967) studied the light transmission and scattering properties of 40μ diameter nylon membrane artificial cells prepared in this laboratory (Chang, 1964; Chang *et al.*, 1966). The optical density was measured in the integrating sphere at four wave lengths, 5400 Å 5600 Å, 5700 Å, and 5800 Å, at a sample depth of 0.071 cm. The hematocrit of the artificial cells was varied from 4% to 44%.

The optical density was plotted against the percentage hematocrit values. As in the case of erythrocytes, the experimental results obtained for artificial cells agreed well with values calculated theoretically.

ELECTRICAL PROPERTIES

We have used a cell electrophoresis apparatus to study the surface properties of artificial cells with different membrane

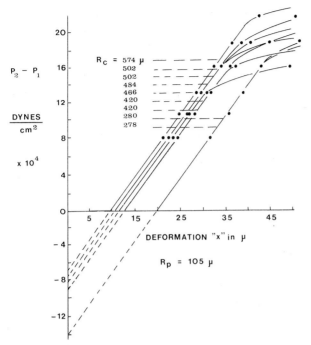

Figure 21. Pressure-deformation curves for various sizes of artificial cells. $R_p = 105\mu$. Note the abrupt change in slope at high pressures. (From Jay and Edwards, 1968. Courtesy of the National Research Council of Canada.)

compositions (Chang, 1965; Chang *et al.*, 1967b). In one set of experiments, it has been observed that whereas the electrophoretic mobilites for erythrocytes or artificial cells with sulfonated nylon membranes are $0.85\mu \pm 0.08\mu$/sec/volt/cm and $1.18\mu \pm 0.24\mu$/sec/volt/cm respectively, artificial cells with nylon membranes do not have any measurable net surface charge. In another set of experiments, the electrophoretic mobilities of dog erythrocytes, of artificial cells with collodion membranes, and of artificial cells with heparin-complex (BHC) collodion membranes are found to be $0.93\mu \pm 19\mu$/sec/volt/cm, $0.93\mu \pm 0.06\mu$/sec/volt/cm, and $0.55\mu \pm 0.11\mu$/sec/volt/cm, respectively, when measured in a phosphate buffer solution. The addition of dog plasma to the buffer solution (1/1000) does not significantly alter the electrophoretic mobility of the dog erythrocytes and of the artificial cells with BHC-collodion membranes ($1.10\mu\pm0.10\mu$ and $0.53\mu\pm0.11\mu$/sec/volt/cm respectively), but lowers the electrophoretic mobility of the artificial cells with collodion membranes to below the measurable limit. Albumin has been incorporated into the microcapsule membranes (Chang, 1969e; Shiba *et al.*, 1970). The electrophoretic mobility of artificial cells with a membrane component containing albumin shows a negative surface charge (Shiba *et al.*, 1970).

Using 3M-KC1 microelectrodes, Jay and Burton (1969) have investigated the potential differences across the membranes of human red blood cells and artificial cells. The nylon membrane artificial cells ($10\mu-500\mu$ diameter) containing hemolysate are prepared as described (Chang, 1964; Chang *et al.*, 1966). In the case of human red blood cells, the potential difference across the cell membranes is -8.0 ± 0.21 mV (SEM), the inside being negative with respect to the outside. When the microelectrode penetrates the artificial cell membrane, a sharp spike is observed. This is followed immediately by a deflection which reaches a maximal negative value of -2 to -4 mV by about five seconds. This then decays in less than ten seconds to a constant value of -0.52 ± 0.02 mV (SEM) (35 measurements). This value remains unchanged until the microelectrode is retracted. As soon as the microelectrode is retracted, the deflection returns to the base line. Jay and Burton suggest that the potential difference of 0.52 ± 0.02 mV (SEM) is due to Donnan equilibrium as a result of the presence of the nondiffusible hemoglobin in the nylon membrane artificial cells.

In further studies, Jay and Sivertz (1969) measure the electrical resistance of membranes of artificial cells ($80\mu-350\mu$ diameter) prepared as described (Chang, 1964; Chang *et al.*, 1966). A suction-pipette holds the artificial cell for impalement by the 3M-KC1 microelectrodes having a tip diameter of 0.5μ. The larger end of the microelectrode is sealed in a 3M-KC1 filled micropipette holder to prevent entry of fluid from the artificial cells. Having measured the electrode resistance, R_e, they impaled the artificial cell membranes and obtained the total resistance of the electrode and nylon membranes, $R_e + R_m$. On withdrawal of the microelectrode, the electrode resistance was checked again. The membrane resistance, R_m, for artificial cells of different diameters was plotted against the reciprocal of the membrane area of each artificial cell, $1/A$. A straight line passing through the origin was obtained. Thus the relationship follows the equation $R_m = px/A$ (p is the resistivity of the membrane and x is the thickness of the membrane, and px is the specific resistivity of the membrane). The specific resistivity was found to be 5.6×10^3 ohm-cm². Using the value of 200 Å for the membrane thickness (Chang *et al.*, 1966), Jay and Sivertz calculated that the resistivity, p, was 3×10 ohm-cm.

Mueller and Rudin (1968) used an internal microelectrode technique to study lipid membrane artificial cells containing red blood cell hemolysate. They found that these membranes have the same electrical properties as the planar bilayer lipid membranes. Pagano and Thompson (1968) studied the transmembrane electrical properties of spherical bilayer membranes several millimeters in diameter. They found that the specific resistance was $3.5-6.8 \times 10^6$ ohm-cm², capacitance was $0.5\mu F/cm^2$, and voltage breakdown potential was 200 mV.

GENERAL PERMEABILITY CHARACTERISTICS

Artificial cells suspended in distilled water present a smooth, spherical surface; but artificial cells placed in a hypertonic solution quickly shrink and show folding of their membranes. This is very striking in the case of nylon membrane artificial cells (Fig. 15), but much less obvious in the case of collodion membrane artificial cells, which have thicker membranes. The shrinkage and the folding are clearly analogous to the phenomena of crenation in erythrocytes and plasmolysis in plant cells. As a result of the osmotic pressure gradient, water is withdrawn from the artificial cells. The shrinkage in Figure 15 exceeds 50 percent of the initial volume as indicated by centrifugation of the suspension in a hematocrit tube. When artificial cells are suspended in a hypertonic solution containing a solute of large molecular or ionic radius, crenation is long-lasting, since penetration of the artificial cells by the solute is slow. But when this solute is of smaller molecular or ionic radius, penetration of the solute is faster and crenation is more quickly reversible, the artificial cells swelling to recover their original form. As would be expected, the duration of crenation is related to the artificial cell diameter, the smaller artificial cells recovering their spherical form more quickly because of the more rapid equilibration of solute across the membrane. There is a satisfactory correlation with molecular weight in the case of the nonelectrolytes tested, and a similar correlation with the hydrated ionic radii in the case of the three alkali metals—Li^+, K^- and Na^+—tested as the chlorides (Fig. 22). Similar studies show that the ionic properties of the carboxyl and amino groups in the nylon membrane of the artificial cells are such that the membrane can become anion-selective or cation-selective by changes in the environmental pH. The permeability characteristics of the artificial cell membranes are further characterized by the more detailed studies discussed in the following sections.

EQUIVALENT PORE RADIUS

According to Van Hoff's equation, a solution with a solute concentration of C can exert an osmotic pressure of π_t where $\pi_t = RTC$. This equation only holds when the osmotic pressure is measured across a membrane which is permeable to water but not to the solute. For a membrane which is permeable to both the water and the solute, the effective osmotic pressure (π_e) is less than the theoretical osmotic pressure (π_t) calculated from this equation. The ratio of the effective osmotic pressure to the theoretical osmotic pressure is the reflective coefficient ($\dfrac{\pi_e}{\pi_t} = \sigma$)

(Staverman, 1957). The reflective coefficient is unity for an impermeant solute, but is less than unity and approaches zero the more permeant the solute. The reflective coefficient is thus a quantitative expression of the permeability characteristics of a membrane, and so much so that experiments demonstrate that the equivalent pore radius of biological membranes can be calculated from reflective coefficients (Solomon, 1961, 1969). We have used this concept to measure the equivalent pore radius of artificial cell membranes (Chang, 1965; Chang and Poznansky, 1968c).

Nylon membrane artificial cells of $270\mu \pm 87\mu$ diameter pre-

NYLON MICROCAPSULES: PERMEABILITY

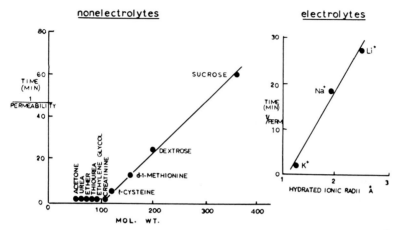

Figure 22. *Left,* permeability of nylon membrane artificial cells to nonelectrolytes. *Right,* permeability of nylon membrane artificial cells to electrolyes: KCl, NaCl, and LiCl. (From Chang, 1965.)

pared by the standard method are dialyzed overnight in 100 volumes of distilled water to remove dialyzable contents. Immediately before use, the artificial cells are washed three times with 10 volumes of distilled water and suspended in 2 volumes of distilled water and kept at 37°C±0.35 degree. At zero time, 0.15 ml of the suspension is added to 1 ml of different test solutions kept at 37°C. A microphotograph is taken at exactly 60 seconds (stopwatch) after the addition of the artificial cells to the test solutions. Observations show that well-prepared nylon membrane artificial cells suspended in water have a smooth membrane and are perfectly spherical, but when they are suspended in solutions of increasing concentrations, folding of the membranes can be observed, somewhat analogous to crenation in erythrocytes or plasmolysis in plant cells. At first this folding occurs in a small proportion of the artificial cells, but as the solute concentration is progressively increased, both the proportion of artificial cells with folded membranes and the degree of folding of the membranes increase. All the artificial cells on each microphotograph can be classified as collapsed or not collapsed,

collapsed artificial cells being defined here as the presence of any observable foldings in the membrane. At least 100 artificial cells are counted for each concentration of solution. Control studies show that counts of the percentage of collapsed artificial cells on the same microphotograph presented to the observer at different times without his being aware give a standard deviation of ±1.13% in five trials. Once collapsed, the artificial cells remain so for at least one minute when four test solutes are used at the stated concentrations. It should be noted at this point that although collapsed artificial cells gradually return to their original spherical shape if the solute can cross the membranes, the recovery of the folded membranes is much slower than would be expected from permeability studies of the undeformed artificial cells. In the folded form, the membranes are no longer stretched by the tension created by the colloid osmotic pressure. The permeability of the folded membranes is thus much less than those of the spherical forms. This makes it possible for the present study to be carried out without a rapid mixing technique.

With glucose as the standard solute, all points fit well on a probit graph by a straight line (Fig. 23). The explanation for this relationship is as follows. The diameters of artificial cells prepared follow a bell-shaped distribution curve with a standard deviation of ±30%. If a curve is made relating $1/R$ (R=radius of artificial cells) and the percentage of artificial cells with diameters of more than a given radius, the results fit well into a straight line when plotted on a probit graph. The slope corresponds quite well with the slope of the relationship between percentage of collapsed artificial cells and concentration of glucose solution. Since nylon membrane artificial cells have internal

→

Figure 23. *Upper*, percentage of collapsed nylon membrane artificial cells as a function of glucose concentration in the medium. Abscissae: concentration of glucose in suspending medium. Ordinates (plotted on probit scale): percentage of artificial cells found to be collapsed in the glucose medium. *Lower*, percentage of collapsed artificial cells in nonelectrolyte solutions plotted on probit scale. ● =sucrose; ▲ = glucose; ○ = propylene glycol; ▼ = ethylene glycol. (From Chang, 1965; Chang and Poznansky, 1968. Courtesy of Interscience Publishers, New York.)

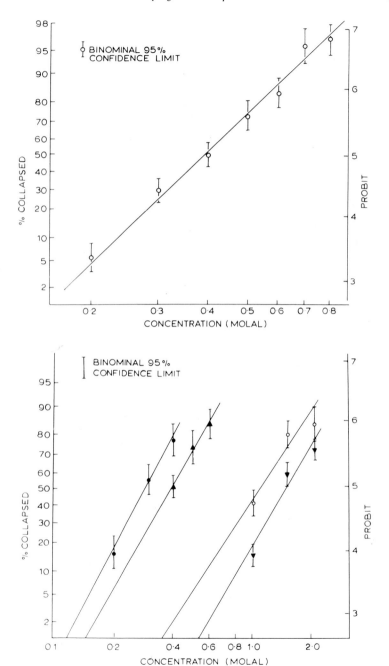

pressures which are inversely proportional to their radii (Jay and Edwards, 1968), the smaller the artificial cells, the greater is the internal hydrostatic pressure, and thus the greater is the external effective osmotic pressure required to collapse the artificial cells (Chang, 1965; Chang and Poznansky, 1968c). The artificial cells are thus microscopic membrane osmometers, each having an internal hydrostatic pressure inversely proportional to its diameter, and each can be collapsed only if the effective external osmotic pressure exceeds that of the internal hydrostatic pressure. The percentage of artificial cells collapsed at a given solute concentration depends on the percentage of artificial cells having an internal pressure which is less than the effective osmotic pressure of the suspending medium. In other words, solutions having the same effective osmotic pressure should cause the same percentage of artificial cells to collapse.

The results obtained from different test solutes are shown in Figure 23. The extrapolation to the x-axis is taken as the threshold concentration. The results obtained are summarized in Figure 23. Each threshold concentration represents the statistically obtained concentration of a given solute which just fails to cause any collapse of artificial cells. As discussed earlier, solutions having the same effective osmotic pressure cause the same percentage of artificial cells to collapse. These statistically obtained threshold concentrations of the four test solutions, sucrose (C_s), dextrose (C_d), ethylene glycol (C_e) and propylene glycol (C_P), thus represent concentrations of these solutes which give the same effective osmotic presure (π_e).

Since $\pi_e = \sigma RTC$, and since the effective osmotic presures (π_e) are the same for the four test solutes, sucrose (s), dextrose (d), ethylene glycol (e), and propylene glycol (p), at their threshold concentrations (C), it follows that

$$\pi_e = \sigma_s.RTC_s = \sigma_d.\ RTC_d = \sigma_e.RTC_e = \sigma_P.RTC_P$$

$$(1)$$

The reflection coefficient is related to the equivalent pore radius (r) of a membrane by the following equation:

$$1-\sigma = \frac{(2(1-a/r)^2-(1-a/r)^4}{2(1-a_w/r)^2-(1-a_w/r)^4}\ \frac{1-2.104a/r+2.09(a/r)^3-0.95(a/r)^5}{1-2.104a_w/r+2.09(a_w/r)^3-0.95(a_w/r)^5}$$

$$(2)$$

where a is the radius of the solute in question and a_w the radius of water. Using a modification of Solomon's principle (1961), the equivalent pore radius of artificial cells is estimated from the results obtained in the following way.

From equation 1 six ratios ($\sigma_s/\sigma_d = C_d/C_s$; $\sigma_s/\sigma_e = C_e/C_s$; $\sigma_s/\sigma_P = C_P/C_s$; $\sigma_d/\sigma_s = C_s/C_d$; $\sigma_d/\sigma_e = C_e/C_d$; $\sigma_d/\sigma_P = C_P/C_d$ and $\sigma_e/\sigma_P = C_P/C_e$) can be obtained. From equation 2, these ratios can be related to the equivalent pore radius as follows:

$$\frac{C_y}{C_x} = \frac{\sigma_x}{\sigma_y} = \frac{1 - \dfrac{2(1-a_x/r)^2 - (1-a_x/r)^4}{2(1-a_w/r)^2 - (1-a_w/r)^4} \cdot \dfrac{1 - 2.104\ a_x/r + 2.09(a_x/r)^3 - 0.95(a_x/r)^5}{1 - 2.104\ a_w/r + 2.09(a_w/r)^3 - 0.95(a_w/r)^5}}{1 - \dfrac{2(1-a_y/r)^2 - (1-a_y/r)^4}{2(1-a_w/r)^2 - (1-a_w/r)^4} \cdot \dfrac{1 - 2.104\ a_y/r + 2.09(a_y/r)^3 - 0.95(a_y/r)^5}{1 - 2.104\ a_w/r + 2.09(a_w/r)^3 - 0.95(a_w/r)^5}}$$

$$(3)$$

where x and y are two nonelectrolytes with known radii a_x and a_y.

Theoretical values for each ratio (C_x/C_y) at different equivalent pore radii can be calculated from equation 3 and plotted (Fig. 24). This way, six theoretical curves are obtained, but of these, two curves (C_e/C_P and C_d/C_s) give no useful information because the slopes of the lines are too flat. For the remaining four pairs, the observed ratios obtained experimentally fall on the curves corresponding to an equivalent pore radius of between 17 Å and 19 Å. Perhaps the best single estimate for the equivalent pore radius is 18 Å.

The equivalent pore radius of the artificial cells prepared by the standard procedure is optimal in serving as a dialyzing membrane which is impermeable to proteins but extremely permeable to small biological molecules. If required, the equivalent pore radius can be varied by varying the time of reaction in the case of nylon membrane artificial cells. Other modifications include the use of membrane materials with different permeability characteristics.

PERMEABILITY CONSTANTS

Because of the rapid equilibration of permeant solutes across artificial cells, a rapid mixing and sampling apparatus is required to study the permeability constants of the enclosing membranes (Chang and Poznansky, 1968c). Nylon membrane

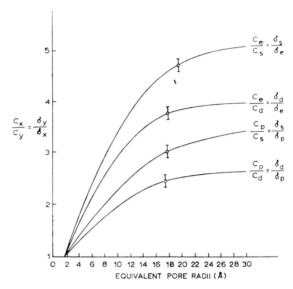

Figure 24. Analysis of equivalent pore radius for nylon membrane artificial cells prepared by the standard procedure. (For explanation see text.) Each of the 4 theoretical curves relates the ratio of reflection coefficients for a given pair of solutes to equivalent pore radii from 0 to 30 Å. The plotted points (mean ± standard deviation) are experimentally determined values for 4 solute pairs. (From Chang and Poznansky, 1968. Courtesy of Interscience Publishers, New York.)

artificial cells with diameters of $210.9\mu \pm 75\mu$ SD are prepared by the method described at a jumbo magnetic stirrer speed setting of 2.5 and Span 85 concentration of 1 percent. Collodion membrane artificial cells with diameters of $214\mu \pm 18\mu$ SD are prepared as described at a jumbo magnetic stirrer speed setting of 2.5, but without the use of Span 85. The artificial cells are finally suspended in a 300mM sodium chloride solution and allowed to stand for 12 hours while the sodium chloride equilibrates completely across the artificial cells. In these experiments, final suspensions contain 110,000 artificial cells per milliliter. A higher concentration of artificial cells is less convenient for rapid mixing and sampling. The results obtained are as shown for nylon membrane artificial cells (Fig. 25 and Table I). Similar results are obtained for collodion membrane artificial cells. In the case of

TABLE I
NYLON MEMBRANE ARTIFICIAL CELLS: PERMEABILITY DATA

Solute	Half-time for Equilibration $T^{\frac{1}{2}}$ (Sec.)	Permeability Constant (P) (cm/sec)	Solute Permeability Coefficient (W) (moles/dyne-sec.)
Urea	4.3	2.01×10^{-4}	8.23×10^{-15}
Creatinine	17.5	0.61×10^{-4}	2.52×10^{-15}
Uric acid	42.5	0.19×10^{-4}	0.77×10^{-15}
Creatine	16.6	0.75×10^{-4}	3.08×10^{-15}
Glucose	26.2	0.54×10^{-4}	2.17×10^{-15}
Sucrose	35.5	0.37×10^{-4}	1.62×10^{-15}
Acetylsalicylic acid	39.0	0.32×10^{-4}	1.31×10^{-15}
Tritiated water (THO)	<1.0	—	—

From Chang and Poznansky (1968c). Courtesy of the Interscience Publishers, New York.

polyurethane membranes, the permeability constants (10^{-5} cm/ min) are 0.32 for CsCl, 0.30 for NaCl, 0.25 for LiCl, 0.31 for NaBr, 0.28 for Na_2SO_4, 0.20 for Na_2CrO_4 (Shigeri and Kondo, 1969).

LIPID MEMBRANES

Spherical Lipid Bilayer Membranes

Mueller and Rudin (1968) combined their technique of planar bimolecular membrane formation (Mueller *et al.,* 1962) with the technique for preparing artificial cells (Chang, 1964) to form "cellules" of about 90μ diameter. Each of these consists of red blood cell hemolysate enveloped in a spherical ultrathin lipid membrane which is 60 to 100 Å thick. When these cellules are exposed to a hypertonic sodium chloride solution, there is a reduction in size within 15 seconds. The reduction in size persists indefinitely, showing good permeability to water but very limited permeability to sodium chloride. Internal microelectrode studies show the same electrical properties as the planar bilayers.

Pagano and Thompson (1967) studied another form of spherical lipid bilayer membranes of much larger diameters of 4 mm. In preliminary studies they found that the water permeability coefficients are 10μ/sec; the specific resistance is 0.35–0.68 ohm–cm^2; capacitance is about 0.5μF/cm^2; breakdown potential is

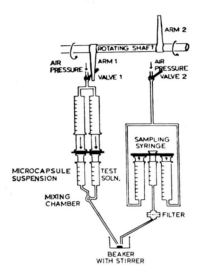

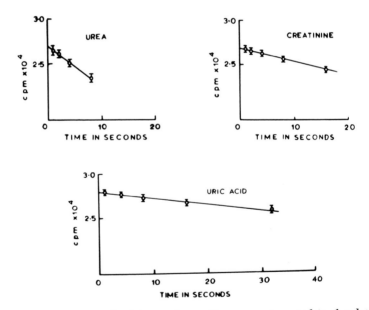

Figure 25. *Upper,* rapid mixing and sampling apparatus used in the determination of permeability constants. Rotating arm (1) opens valve (1) to air pressure and triggers the rapid injection of artificial cell suspension and test solution through the mixing chamber into a collecting beaker. After a

→

200 mV. Thus the properties are very similar to those of the planar system. In further detailed studies Pagano and Thompson (1968) studied the isotopic fluxes of ^{22}Na and ^{36}Cl and found that the unidirectional fluxes M_{Na} and M_{Cl} were respectively 0.39 $\pm$ 0.02 and 90.2 $\pm$ 0.8 pmoles/cm^2/sec at 30°C. The chloride flux was also studied as a function of temperature, and the activation energy for chloride permeation across the bilayer was calculated to be 10.7 $\pm$ 0.4 kcal/mole. Other studies were made by Pagano and Thompson to compare the experimental results of the sodium and chloride flux measurements with the theoretical values as calculated from the electrical parameters of the spherical bilayer membrane. It was found that while the sodium flux falls within the range of values calculated from the electrical data (0.0298−1.194 pmoles/cm^2/ses); the chloride flux is of much greater magnitude than the theoretical value (0.061−0.245 pmoles/cm^2/sec). In fact, the chloride flux is more than 300 times greater than the calculated value. The authors postulated that the chloride flux consists of an ionic component to account for the electrical parameters, and an exchange diffusion component. They then carried out experiments to examine M_{Cl} as a function of chloride ion concentration within the spherical bilayer. The chloride influx is very small when no chloride ion is in the interior aqueous phase. They found that the addition of 0.05M NaCl to the internal aqueous phase increases the observed chloride flux markedly. In addition, the chloride flux exhibits a saturation kinetic mechanism reaching a maximal rate by 0.1M NaCl internal concentration. All these observations have led the authors to propose that a carrier mechanism is present for chloride permeation acros the bilayer spherical membranes. Pagano and Thompson (1968) suggested the following as possible carriers:

---

predetermined interval, rotating arm (2) of kymograph triggers the sampling syringe to aspirate an aliquot of artificial cell free supernatant through the filter. *Lower*, permeability constants, example of results obtained. Rate of entry of solute measured as the rate of decrease of radioactivity in the suspending medium. Concentration of artificial cells (diameter 210.9$\mu\pm$ 75.0μ) in the suspension after the addition of an equal volume of the test solution is 55,000/ml. (From Chang and Poznansky, 1968. Courtesy of Interscience Publishers, New York.)

(1) sodium ion, (2) a small unknown molecule present in the membrane, (3) the phospholipid, or (4) a heavy metal phospholipid complex.

Lipid-Macromolecular Complex Membranes

The exact molecular organization of biological cell membranes is not known. However, that biological membranes are not composed of lipids alone is without dispute. Recently, Korn (1968) proposed a possible model in which he suggested that in the biosynthesis of biological cell membranes the protein unit of the membrane is formed first, followed by the addition of the lipid components. If this theory is correct, then it would be of experimental interest to form a spherical ultrathin membrane of polymer or cross-linked protein, then coat this with natural lipids. This has been done (Fig. 12) (Chang, 1969b), and experiments have ben carried out on these model systems (Fig. 26). In a 150 mM NaCl medium, sodium influx across uncoated spherical ultrathin polymer membranes is about 6×10^{-5}mM/hr-mm^2. Coating with an equimolar lecithin-cholesterol mixture reduces the sodium influx to about 4×10^{-6} mM/hr-mm^2. This value obtained for lipid-coated artificial cells is comparable to the value obtained for red blood cell membranes. It is interesting to note that in the membranes formed by lipids alone, the sodium influx is many orders of magnitude lower than that of biological cell membranes or lipid-coated artificial cell membranes. Further detailed studies have been carried out (Rosenthal and Chang, 1971) using rubidium. These lipid-coated artificial cell membranes do not exhibit any significant degrees of selectivity between sodium and rubidium. On the other hand, in the presence of Valinomycin, the rubidium influx is enhanced to a much greater extent than the sodium influx. Thus, the permeability coefficients of control nylon membrane artificial cells and of lipid-coated nylon membrane artificial cells to rubidium are respectively $19.5 \pm 2.1 \times 10^{-6}$ cm/sec and $1.45 \pm$ SD 0.56×10^{-6} cm/sec. Valinomycin at a concentration of 5×10^{-6} M significantly increases the permeability coefficient of the lipid-coated microcapsules to $4.91 \pm 1.60 \times 10^{-6}$ cm/sec. In the same studies (Rosenthal and Chang, 1971) it has been demonstrated that the lipid-coating reported (Chang,

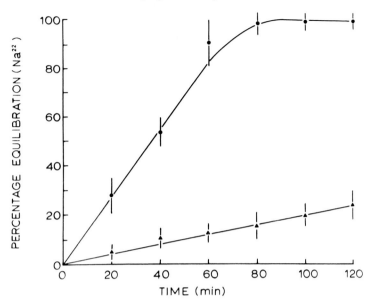

Figure 26. Rate of equilibration of ^{22}Na across artificial cells. Control nylon membrane artificial cells (●) compared to lipid-coated nylon membrane artificial cells (▲). (From Chang, 1969b.)

1969b) corresponds to an ultrathin coating of lipid molecules on the artificial cells.

DIFFERENTIAL DIALYSIS

The different permeability of artificial cells for substances of different molecular size suggests their possible use in differential dialysis. Solutes of large molecular size that cannot penetrate the artificial cells are excluded and emerge from the column without retardation, while solutes capable of diffusing into the interior are retarded. Figure 27 shows that a column of artificial cells can separate hemoglobin from glucose almost completely in a single passage (Chang, 1965). The characteristics of the column are given in the legend. In other tests partial separation was obtained of glucose and sucrose, and of sucrose and urea; obviously, the size, permeability, and flow rate of the solute molecules are important parameters. Although this is in theory somewhat similar to those employed in chromatography using Sephadex

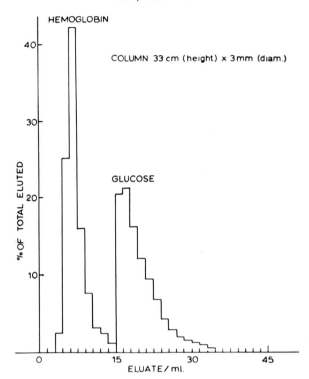

Figure 27. Column with dimensions as shown, packed with nylon membrane cells of mean diameter about 90μ, prepared by the standard method using 1% (v/v) Span 85 and jumbo magnetic stirrer speed setting of 1. Flow rate 0.09 ml/min; eluate collected in 0.2 ml fractions. (From Chang, 1965.)

granules, the solid polymer matrix of the Sephadex granules cannot be expected to have anything close to the high permeability characteristics of the ultrathin membrane of artificial cells. Furthermore, the artificial cell contents can modify permeant materials entering the artificial cells.

Craig (1964) discussed differential dialysis using hollow tubular membranes. Here, the speed of diffusion equilibrim was secured by the use of the tubular membranes separating two aqueous phases in counter-current flow. Recent analysis (Flinn and Cherry, 1970) demonstrated the efficiency of artificial cells over a tubular system of membranes. The smallest useful hollow fibers have a diameter of 50μ with a wall thickness of 10μ. A much thinner membrane is possible with microscopic spherical membranes like artificial cells. Thus 4600 spherical membranes 3.45μ in diameter and having a wall thickness of 0.5μ would equal the core volume of the hollow fiber section. The rate of material transport across the total membrane surface of this number of spheres is nearly 400 times that for the hollow fiber. For artificial cells of 0.02μ wall thickness as prepared in this laboratory (Chang, *et al.*, 1966), the rate would even be greater. The main advantage of artificial cells is the combination of high surface area to volume relationship and ultrathin membranes. No other membrane systems can at present have a comparable efficiency.

DISCUSSION

From the results obtained so far, artificial cells possess some of the simpler biophysical properties of biological cells. Of special importance are their permeability characteristics. The results obtained are as shown (Figs. 22, 24, 25, and Table I).

The equivalent pore radius of 18 Å for the membranes of nylon artificial cells is midway between that for human erythrocyte membranes (4.2 Å) and glomerular capillary walls (35.5 Å). Thus although molecules of small molecular dimensions can readily cross the membranes, macromolecules like ovalbumin (molecular radius of 28.5 Å), hemoglobin (molecular radius of 32.5 Å), and a number of enzymes studied, cannot leak out of the artificial cells in which they have been enclosed (Fig. 28). The equivalent pore radius of 18 Å is slightly smaller than that of cellulose membranes; however, the much thinner nylon membrane (200 Å) results in a permeability constant P for urea of 2×10^{-4} cm/sec, and the large surface area to volume relationship of the artificial cells results in an extremely rapid equilibration of solute (e.g. for urea, half-time of 4.3 seconds across artificial cells of 207μ mean diameter). Preliminary experiments indicate that, as would be expected, the half-time for equilibration is reduced even further when artificial cells of smaller diameters are tested. The permeability properties of such enclosing polymer membranes thus fulfill most of the requirements for artificial cells described in the introductory section. Thus protein,

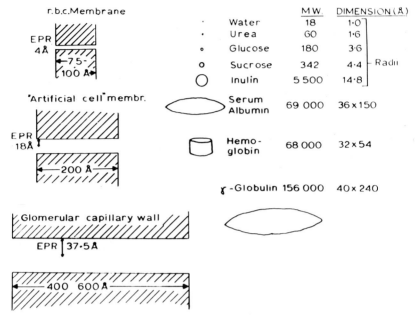

Figure 28. Schematic representation comparing equivalent pore radius (EPR) to solute molecules.

enzyme, or suspension enclosed within these artificial cells would not leak out, but small molecules like substrates or toxins can equilibrate rapidly across the artificial cell membranes. Permeability characteristics of artificial cells can be modified by lipid coating, charge, carrier molecules, and others.

Chapter 4

BIOLOGICAL ACTIVITY OF
ARTIFICIAL CELLS

INTRODUCTION

H AVING characterized some of the biophysical properties of artificial cells, we are now in a position to examine their biological properties. Whether in solution or imbedded in membranes of intracellular organelles, enzymes in nature are located most abundantly within biological cells. While retained within the cells and protected from extracellular environments, enzymes act efficiently on permeant substrates entering through the cell membranes by diffusion or special transport mechanisms.

Initially, I was interested in studying an artificial system of cellular dimensions in which individual microdroplets of erythrocyte hemolysate with its hemoglobin and complex enzyme systems are enveloped in spherical ultrathin polymer membranes (Fig. 2) (Chang, 1957). In this way, the enveloped proteins are retained within cell-like microscopic dialysis bags and prevented from coming into direct contact with the extracellular environment. Permeant external molecules, like oxygen and carbon dioxide, can diffuse rapidly across the enclosing membrane to interact with the enclosed erythrocyte hemolysate. Additional enzyme systems can also be added to the red blood cell hemolysate to be enclosed in artificial cells. Thus enzymes could be enclosed in artificial cells by first dissolving or suspending them in the hemolysate solution and then proceeding as described for the preparation of artificial cells. It should be emphasized that the preparation of stable artificial cells is facilitated by the presence of a high concentration of protein in the aqueous phase. A small fraction of the protein takes part in the formation of the cell membrane by cross-linking or by coacervation. The osmotic pressure of the colloid also helps to retain the turgor of the artificial cell membranes, which would otherwise tend to collapse in an aqueous environment. Thus, dilute enzyme preparations can be

most conveniently enclosed in artificial cells if they are first added
to the hemoglobin solution or hemolysate. Only in a very few
cases can enzyme solutions (e.g. Sigma type V urease 0.15 gm/
ml) be microencapsulated inside artificial cells without the addition of further protein. In these cases, the membranes are usually
not well formed, even though a high concentration of the enzymes
has been used. Another point which should be emphasized is that
the chemicals used in the interfacial polymerization procedure
may inactivate some types of enzymes, e.g. catalase and uricase,
but have no marked effects on other types of enzymes, e.g. urease
and asparaginase. The interfacial precipitation procedure, on the
other hand, did not have any major adverse effect on all the
enzymes tested so far, for instance, carbonic anhydrase, catalase,
trypsin, uricase, urease, and asparaginase. However, in the case
of the interfacial precipitation procedure, the physicochemical
properties of the solution to be encapsulated are extremely important. One of these is the pH of the solution. For example,
when 1.8 gm of urease (Sigma type II powder, 1720 units/gm)
was dissolved in 10 ml of hemolysate containing tris buffer
(0.08M), the high acidic content of Sigma type III urease lowered the pH of the final solution to 6.2. The collodion membrane
artificial cells prepared from this solution were very fragile and
easily ruptured. On the other hand, if the pH was maintained at
9 by the use of a high concentration of tris buffer (0.48M), the
artificial cells prepared had much stronger membranes.

Besides enzymes and hemolysate, cell extracts, cell homogenates, or insolubilized enzymes have been added to the hemolysate and then enclosed in the artificial cells. Some of the enzymes
and proteins enclosed in artificial cells are summarized in Table
II. Enzymes can be microencapsulated, then stabilised by treatment with glutaraldehyde (Chang. 1971d).

MODEL SYSTEM

Artificial Cells Containing Urease

Urease, which catalyzes the hydrolysis of urea, is a globulin
with a molecular weight of 480,000. It is unstable in solution,
losing 50 percent of its activity in 24 hours at 37°C, but is more
stable in the presence of 2 percent gum arabic or 5 percent egg

TABLE II

ENZYMES AND PROTEINS ENCLOSED IN ARTIFICIAL CELLS

1. Hemoglobin and enzymes of erythrocyte hemolysates (Chang, 1957, 1964, 1965; Chang *et al.*, 1963, 1966; Toyoda, 1966; Jay and Edwards, 1968; Jay and Burton, 1969; Jay and Sivertz, 1969; Mueller and Rudin, 1968; Kitajima *et al.*, 1969; Sparks, *et al.*, 1969).
2. Urease (Chang *et al.*, 1963; Chang and MacIntosh, 1964; Chang, 1964, 1965, 1966, 1969d; Chang *et al.*, 1967b; Chang and Loa, 1970; Levine and LaCourse, 1967; Sparks *et al.*, 1969; Falb *et al.*, 1968; Kitajima *et al.*, 1969).
3. Carbonic anhydrase (Chang, 1964, 1965).
4. Uricase (Chang, 1964, 1965; Chang *et al.*, 1966).
5. Trypsin (Chang, 1964, 1965).
6. Catalase (Chang, 1967; Chang and Poznansky, 1968; Kitajima *et al.*, 1969).
7. Asparaginase (Chang *et al.*, 1968; Chang, 1969c, 1969d, 1969f, 1971b).
8. Albumin (Chang, 1964, 1965; Chang *et al.*, 1966; Shiba *et al.*, 1969).
9. Lipase (Kitajima *et al.*, 1969).
10. α-glucosidase (Ryman, 1968).

albumin and very stable in the insolubilized form. Its substrate, urea, is the major diffusible nitrogenous constituent of the body fluid, and the interesting possibility of a demonstration of an *in vivo* action of urease enclosed in artificial cells has prompted the investigation of the efficiency of this system (Chang, 1965).

Urease dissolved in hemolysate was enclosed in nylon membrane artificial cells by the methods described in the previous section. After washing with phosphate buffer to remove any broken artificial cells, urease activities were analysed by the procedure of Van Slyke and Archibald (1944), as the rate of rise of pH of a buffered urea solution. The result is shown in Figure 29. Five Sumner units of the enzyme enclosed in artificial cells had an activity corresponding to that of 1.83 ± 0.10 Sumner units of enzyme in free solution. Thus the activity of the artificial cell urease was about 37 percent of the activity of the same amount of enzyme in free solution. It also shows that a sample of buffer in contact with its own volume of urease-loaded artificial cells for 12 hours acquires no measurable enzymatic activity, showing that the enzyme does not leak to a significant degree out of the stored nylon membrane artificial cells (Chang, 1965). In another set of experiments, nylon membrane artificial cells containing urease were ruptured in a cell homogenator (glass-pistal). The

Artificial Cells

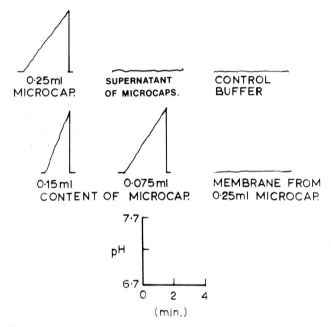

Figure 29. Urease activity, measured as rate of rise of pH of urea-buffer medium. No pH change in the presence of supernatant from a 50% suspension of artificial cells stored for 12 hours. The artificial cells are homogenized and separated into the membrane fraction and homogenate fraction. Nearly all the enzyme activity is located in the membrane-free homegenate.

homogenate was recovered and the membranes washed three times with phosphate buffer. No measurable urease activity was found in the isolated artificial cell membranes. The urease activity was all in the homogenates. Since the enclosed urease did not leak out of the artificial cells and since the membranes of the artificial cells did not contain any measurable enzyme activity, it appears that in order for the urea to be acted on by the enclosed enzyme, it has to diffuse into the artificial cells. Further experiments were done to test the effect of artificial cell diameter on the activity of the enclosed enzyme. In the case of the smaller artificial cells $(27.1\mu\pm11.9\mu)$, the rate of enzymatic conversion was about 39 percent of the same amount of enzyme activity in free solution. In the case of the larger artificial cells $(89.8\mu\pm 26.0\mu)$, the activity was about 21 percent of the activity in free solution.

Additional experiments were done to test the stability of the urease enclosed in artificial cells (Chang, 1965). Urease encapsulated alone without the presence of hemolysate lost its activity rapidly with a half-time of three hours when stored at 37°C. When urease was enclosed together with hemolysate in artificial cells, the stability of the enclosed enzyme was greatly improved. This is shown in Figure 30. It was found that the half-time of the hemolysate-stabilized microencapsulated urease was about one week when kept at 37°C and two weeks when kept at 4°C. Urease enveloped in collodion membrane artificial cells or BHC-collodion membrane artificial cells also showed similar *in vitro* activities.

Site of *in vivo* Introduction

In order to study the *in vivo* action of artificial cells, a site has to be selected for their *in vivo* introduction. Except for blood cells, other cells are bathed in interstitial fluid, an ultrafiltrate of blood plasma. Peritoneal fluid is an ultrafiltrate of blood plasma (Maurer, *et al.*, 1940). Small molecular weight substances exchange rapidly between the peritoneal blood vessels and the peritoneal fluid in the peritoneal cavity (Courtice and Simmonds, 1954). In addition to the demonstrated efficient exchanges between blood plasma and peritoneal fluid, the peritoneal cavity has the potential capacity to hold a large volume of artificial cells. Thus it appears that the peritoneal cavity would be an ideal experimental site for the *in vivo* introduction of artificial cells (Chang, 1965).

Effects And Fates Of Injected Artificial Cells

Before the *in vivo* action of artificial cells could be investigated, a knowledge of the toxicity and fate of intraperitoneally injected artificial cells would assist in the interpretation of the results. Nylon membrane artificial cells were chosen for these studies for

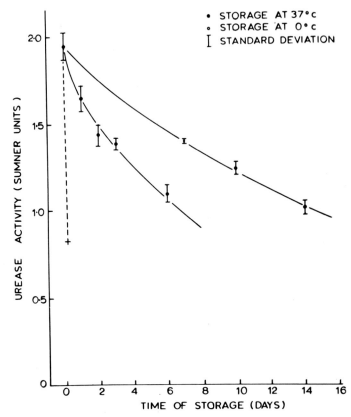

Figure 30. Stability of urease enclosed in nylon membrane artificial cells, stored in phosphate buffer at pH 6.7. Continuous lines represent activity after storage at 0°C (upper line) and at 37°C (lower line) for enzyme encapsulated with hemolysate. Discontinuous line represents activity after storage at 37°C for enzyme encapsulated without hemolysate. (From Chang, 1965.)

the following reason: Teflon® and silicone rubber, though extremely inert biologically, are not semipermeable to water or to aqueous solutes. Nylon, though it causes some tissue reactions in the sheet or block form, is nearly inert when used as a suture material.

Since particles the size of erythrocytes are removed from the peritoneal cavity, we have used larger nylon membrane artificial cells of 100µ mean diameter prepared as described. After washing to remove Tween 20, the enclosed hemoglobin was tagged with sodium chromate Cr 51. Male rats of average weight 130 gm were divided into six groups. The animals of five groups were each given 1 ml intraperitoneally of the 25 percent suspension of nylon membrane artificial cells. Each rat of the sixth group received 1 ml of the supernatant from the suspension. All the animals remained active and healthy, without significant changes in weight, abdominal tenderness or rigidity, or alteration of bowel movements. On opening the abdomens, there were no signs of inflammation in any abdominal structure, except for one case in which all the artificial cells had been injected accidentally into a small pocket of omentum resulting in some local inflammation and fibrosis. In all other animals, larger individual nylon membrane artificial cells and small aggregates could be seen loosely sticking to the abdominal wall or omentum and could easily be picked up with fine forceps, leaving no signs of fibrosis or inflammation at the sites. After the first week, the artificial cells were well dispersed over the whole peritoneal cavity, but at the second week and thereafter there was some tendency for them to be found in larger numbers in the upper part of the peritoneal cavity, where many of them were located at the upper surface of the liver and spleen just below the diaphragm. None were observed on the peritoneal surface of the diaphragm. No significant radioactivity was detected in the lung, liver, spleen, lymph nodes, blood, or saline wash fluid. Recovered radioactivity was associated with the artificial cells or small aggregates of artificial cells which stuck loosely to the abdominal structures.

A dog was given intraperitoneal injections of nylon membrane artificial cells at weekly intervals for three weeks. It was then followed for one year. The dog continued to be active, gained weight, and showed no signs of abdominal tenderness or rigidity. Examination after one year showed no fibrosis or chronic inflammation in the peritoneal cavity.

in vivo Action Of Artificial Cells Containing Enzymes

Artificial cells containing urease were tested (Chang and MacIntosh, 1964b, Chang, 1965). The principle is shown diagrammatically in Figure 31. The enzyme urease does not leak out from

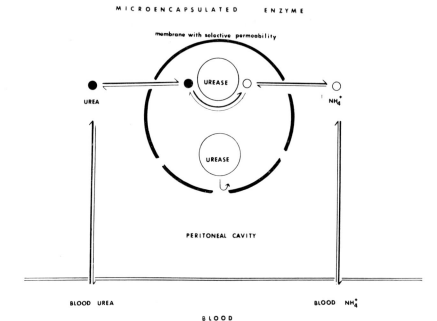

MICROENCAPSULATED   ENZYME

membrane with selective permeability

UREASE

UREA

NH₄⁺

UREASE

PERITONEAL CAVITY

BLOOD UREA

BLOOD NH₄⁺

BLOOD

Figure 31. Schematic representation of the action of intraperitonally injected artificial cells loaded with urease. (From Chang, 1965.)

the artificial cells. The size of the artificial cells (100μ average diameter) prevents the artificial cells from leaving the peritoneal cavity. Thus, for the urease in the artificial cells to have significant *in vivo* activity, its substrate, urea, would have to diffuse into the peritoneal cavity from the blood and then across the artificial cell membrane to be acted upon by the enclosed enzyme. The product, ammonium carbonate, would have to diffuse in the reverse direction to raise the blood ammonia level. Thus, the *in vivo* activity of the enclosed enzyme could be assessed by following the blood ammonia levels.

The results obtained from anesthetized dogs are shown in Figure 32. Control artificial cells containing red blood cell hemolysate (0.25 ml/kg) injected intraperitoneally produced no significant changes in the arterial blood levels of the animals. However, after the intraperitoneal injection of artificial cells containing both hemolysate and urease (100 Sumner units in 0.25 ml/kg),

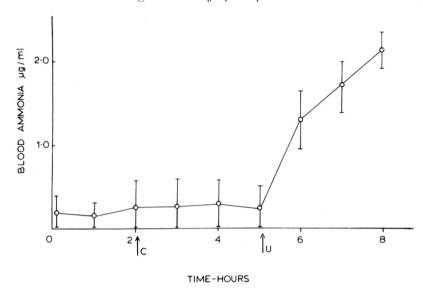

Figure 32. Effect of urease-loaded artificial cells on blood ammonia in dogs anesthetized with Nembutal; the graph summarizes data from 3 experiments (mean ± SD). C = injection of control artificial cells (no urease, 0.25 ml/kg). U = injection of urease-loaded artificial cells (0.25 ml and 100 Sumner units/kg). (From Chang, 1965.)

the arterial blood ammonia levels rose. At the end of the experiments, three to four hours after the injection of the artificial cells containing urease, the blood amonia level was still rising. No significant changes were observed in blood pressure, electrocardiogram, or respiration. As a test for leakage of protein from the injected artificial cells, the enclosed hemoglobin was tagged with Cr 51. No radioactivity was detectable in the circulating blood.

The changes in blood ammonia obtained in an unanesthetized dog are shown in Figure 33. The intraperitoneal injection of the control artificial cells had no significant effect on either the blood ammonia level or the general state of the dog. When the artificial cells containing urease were injected (0.5 ml containing 50 Sumner units/kg), there was likewise no immediate effect on the behavior of the animal. Later, corresponding to the peak of the blood ammonia level, the animal became more sedated and

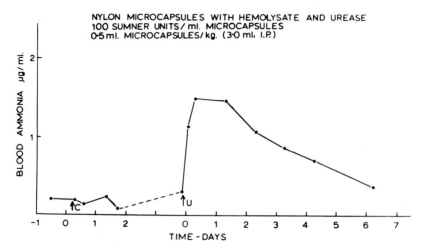

Figure 33. Effect of artificial cells on blood ammonia in an unanesthetized dog. C= control artificial cells (0.5 ml/kg). U = urease-loaded artificial cells (0.5 ml and 50 Sumner units/kg). In this test, the urease preparation (NBC) was stabilized by the addition of hemolysate. (From Chang, 1965.)

walked with an unsteady gait. Twenty-four hours later, even though the blood ammonia level remained high, the animal appeared to be perfectly normal and has remained so up to one year after the injection. It is seen that the *in vivo* activity of the urease in the artificial cells, as indicated by the blood ammonia level, has declined slowly with a half-time of two to three days. This decrease in activity is not entirely related to the decrease in enzyme activity when stored at 37°C (half-time of about seven days); it could be due to the coating of the artificial cells by protein, fibrin, or phagocytes, thus reducing the permeability of substrates across the membrane.

The results of these experiments show conclusively that urease in the artificial cells when injected intraperitoneally can act efficiently on endogenous urea, converting it into ammonia. Only a small proportion of the body's urea was changed to ammonia in these tests, but it must be remembered that the intact liver was simultaneously reconverting ammonia to urea. In addition, the *in vivo* action of urease appeared to rise steeply with enzyme dosage. The greater effectiveness of the higher dosage indicates that the rate-limiting factor for ammonia formation under the conditions of these tests was the amount of enzyme present in the peritoneal cavity rather than the rate of transfer of urea or ammonia across the peritoneal membrane.

These results led us to suggest that artificial cells containing enzymes might be useful for enzyme replacement therapy (Chang, 1964, 1965; Chang and MacIntosh, 1964b). In this form, the enzyme, while acting on external permeant substrate, cannot leak out of the artificial cells to become involved in allergic or immunological reactions (Fig. 2).

Chapter 5

EXPERIMENTAL ENZYME THERAPY
INTRODUCTION

R APID progress has been made in the field of enzymology. In the face of all this progress, enzymes, despite their fundamental importance, still have not gained their rightful place as one of the important tools in clinical therapy. There are many problems which limit the therapeutic use of enzymes. Unless they are highly purified, the contaminants may give rise to unwanted reactions. Since, at present, substantial amounts of enzymes are more plentiful in the form of foreign proteins obtained from heterogeneous sources, there may be problems associated with hypersensitivity reactions, immunological reactions, and rapid removal and inactivation. Further, there is the problem of keeping the enzymes in the body at the sites where they can function efficiently. There are a few situations where these problems are not as important: when digestive enzymes are used orally to supplement deficient exocrine pancreatic secretions; when enzymes are given by external application; when highly purified enzymes are used for short periods of time; or when enzymes from homologous sources are used. Unfortunately, in most other situations these problems prevent the therapeutic use of enzymes. However, in recent years it has been shown that with further development there is the possibility that some of these problems might be circumvented (Chang, 1964, 1969d; Chang and MacIntosh, 1964b; Chang and Poznansky, 1968a). In nature, most of the enzymes act while remaining in an intracellular environment. Thus, the possible uses of enzyme-loaded artificial cells for enzyme therapy has been investigated experimentally in two typical model systems, one involving enzyme replacement for congenital enzyme deficiency diseases; another involving enzyme therapy for the suppression of substrate dependent tumours. This chapter gives an outline of the results obtained.

ARTIFICIAL CELLS FOR INBORN ERRORS OF METABOLISM

Introduction

Since Garrod's introduction (1909) of the concept or inborn errors of metabolism, an increasing number of diseases have been added to this group. Many of these have been shown to be due to specific enzyme defects. Enzyme defects need not necessarily imply the absence of the enzymes. It may be due to a structural genetic mutation at the reactive site, resulting in an enzymatically inactive molecule. In other cases, normal enzymes may be present, but cannot act, either because of the presence of an inhibitor or because of the deficiency in cofactors. In all cases, the enzyme defects result in either an accumulation of substrates to a toxic level or the deficiency of essential products (Fig. 34). At present, clinical treatment of these diseases is by "environmental engineering" (cf. Scriver, 1969). This includes restricting the intake of the substrate (e.g. dietary restriction), replacing the deficient product (e.g. hormonal replacement), or supplementing the cofactor (e.g. vitamin B_6 administration). Even if the enzymes which are deficient in "inborn errors of metabolism" become available, they will have to be administered in a form which can reach the relevant sites of action without being rapidly inactivated and removed and without involving immunological and hypersensitivity recations. One line of research in this laboratory (Chang and Poznansky, 1968a) involves the use of artificial cells for experimental replacement therapy in mice with a congenital

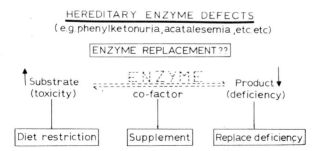

Figure 34. Schematic representation of the available approaches in the treatment of enzyme deficiency diseases.

deficiency in the enzyme catalase. These acatalesemic mice have been recently developed by Feinstein *et al.* (1966a). One of the mutant strains, C_s^b, has a blood catalase activity about 2 percent of normal and a total body catalase activity 20 percent of normal. This is an example of enzyme polymorphism (Feinstein *et al.,* 1968) since, despite the lack of catalase activity in the blood, the mutated catalase molecules are still present in an inactive form. Using this strain of mice they (Feinstein *et al.,* 1966b) showed that these mice are ideal for the investigation of enzyme replacement therapy. For example, they showed that injections of catalase solution protected the animals from hydrogen peroxides, whereas injections of a form of insoluble catalase derivative did not offer any significant protection. Repeated injections of beef catalase produced antibodies to the enzyme. Since catalase has been microencapsulated in collodion membrane artificial cells (Chang, 1967), we started investigations to study the various aspects of enzyme replacement in these acatalasemic mice C_s^b using artificial cells containing beef catalase (Chang and Poznansky, 1968a; Poznansky and Chang, 1969; Poznansky, 1970; Chang, 1971c).

Preparation

As discussed under the section of artificial cells containing cell homogenate, catalase of red blood cell hemolysate enclosed in artificial cells retains its activity in the case of collodion membrane artificial cells. Further studies were done in this laboratory to artificially increase the concentration of catalase in the hemolysate of the artificial cells (Chang, 1967; Chang and Poznansky, 1968a). This was done by dissolving crystalline lyophilized catalase (Nutritional Biochemical Co.) 500 mg in 10 ml of tris-buffered (0.495M 2-amino-2-hydroxymethyl-1,3-propanol) 10 gm% hemoglobin (Worthington Co.) solution. The final solution was filtered through No. 42 Whatman filter paper to remove any undissolved particles. Three milliliters of the filtrate were enclosed in collodion membrane artificial cells by the procedure already described for preparing 80μ diameter artificial cells. The interfacial polymerization procedure that has been described for nylon membrane artificial cells was unsuitable for the preparation of catalase-loaded artificial cells, because it inactivated the enclosed catalase. Before use, the artificial cells were washed at least six times with 10 volumes of saline or until there was no visible leakage of hemoglobin from the small proportion of imperfect artificial cells.

in vitro **Studies**

Sodium perborate is one of the substrates used for assaying the enzymatic activity of catalase. *In vitro*, microencapsulated catalase acted on this substrate at a rate 25 percent as efficient as that of the same amount of enzyme in free solution. No detectable catalase leaked out of the artificial cells during the test period, but sodium perborate equilibrated across the artificial cell membrane with a half-time of about 18 seconds. It was also noted that the erythrocyte hemolysate in artificial cells containing added catalase retained the red color of oxyhemoglobin, whereas when no catalase was added, a higher proportion of the enclosed hemoglobin was in the form of methemoglobin. *In vitro* studies on polystyrene artificial cells containing either hemolysate or catalase done in another laboratory (Kondo, 1968) show that the enclosed catalase also acted efficiently on its substrate, hydrogen peroxide. They found that the reaction rate was proportional to the amount of artificial cells containing the enzyme.

in vivo **Studies**

In vivo experiments were performed in this laboratory (Chang and Poznansky, 1968a) using the Feinstein strain of acatalasemic mice C_s^b and the corresponding normal strain C_s^a. In one set of experiments, one group of normal mice and three groups of acatalasemic mice were used (5 mice per group). Each mouse received a subcutaneous injection of 0.014 mM/gm body weight of sodium perborate (0.1M aqueous solution) as substrate for catalase. As controls, one group of acatalasemic mice and the group of normal mice received no other treatment. In the second group of acatalasemic mice, each animal had been given an intraperitoneal injection of 0.15 ml/gm body weight of a 50 percent suspension of artificial cells loaded with catalase just before receiving the perborate injection. In the third acatalasemic group, each mouse received just prior to the subcutaneous injection of perborate 0.75 mg/gm of liquid catalase intraperitoneally.

It was observed that in acatalasemic mice which were given an injection of sodium perborate without the previous protection of catalase, the red color of the pupils turned brown and the animals became increasingly immobile, so that after 20 minutes many of them were flaccid and in respiratory distress. Acatalasemic mice protected by the intraperitoneal injection of microencapsulated catalase responded differently to the subcutaneous injection of perborate. Five minutes after the injections they became slightly immobile and the color of the pupils darkened; in 10 minutes, however, the mice became more active, the pupils returned to their normal color, and, except for one mouse, the animals were moving about freely at the end of the 20 minutes. Acatalasemic mice injected with catalase solution and normal mice recovered even faster.

Exactly 20 minutes after the injection of perborate, each animal was homogenized for two minutes in four times its weight of 1N sulfuric acid. Three milliliters of the homogenate was added to 6 ml of solution containing sulfuric acid (1N) and trichloroacetic acid (5%). This suspension was filtered and 6 ml of filtrate was titrated with potassium permanganate (0.005N). All measurements were performed in triplicate. For these measurements, two controls were used. First, titratable material in the homogenate other than sodium perborate was measured in normal and acatalasemic mice which had not received any injected perborate. This value was used as a blank in the calculation of total body sodium perborate. In another control measurement sodium perborate was added to the homogenate, the catalase of which had been inactivated by the presence of sulfuric acid (1N). When the filtrate of this homogenate was titrated, more than 95 percent of the perborate added could be titrated by potassium permanganate (0.005N).

The results of the *in vivo* experiments showed that when the total body perborate was measured in control animals 20 minutes after the injection of sodium perborate, only 2.5 percent±2.5 percent (SD) of the injected substrate could be recovered from the normal mice, whereas 70 percent±8.2 percent (SD) could be recovered from the acatalasemic mice. When the acatalasemic mice had received injections of liquid catalase or catalase-loaded artificial cells just before the injection of the sodium perborate, only 7 percent±3.6 percent (SD) and 16 percent±3.5 percent (SD) of the perborate could be recovered from the respective groups. Further detailed studies using a more sensitive method of analysing perborate have been carried out (Poznansky and Chang, 1969) and are being prepared for publication. Recently, more precise and quantitative values of perborate recovery were obtained for the control and treated acatalasemic mice (Fig. 35)

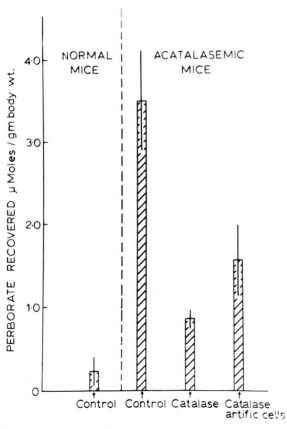

Figure 35. Injected sodium perborate recovered from normal C^a$_s$ and acatalasemic C^b$_s$ mice 20 minutes after injection. (From Chang, 1971b.)

(Chang, 1971b). These results again supported the earlier preliminary findings that artificial cells containing catalase acted
efficiently in replacing the enzyme deficiency.

Thus intraperitoneally injected liquid catalase and artificial
cells containing catalase both acted efficiently in replacing the
deficient enzyme. However, 20 minutes after the intraperitoneal
injection of catalase-loaded artificial cells, there was no increase
in blood catalase level; on the other hand, 20 minutes after the
intraperitoneal injection of catalase in free solution, there was a
significant increase in the blood catalase level. Thus intraperitoneally injected catalase-loaded artificial cells, unlike catalase in
free solution, acted efficiently *in vivo* without leaking out of the
artificial cells. This is an important difference, because repeated
injections of free enzymes like catalase from heterogeneous
sources induce the formation of antibodies in the recipient. In the
case of enzyme-loaded artificial cells, our earlier and present
studies show that enclosed enzymes do not leak out but still act
efficiently on external permeant substrates. Furthermore, gamma
globulins cannot cross the artificial cell membrane because of the
small equivalent pore radius. Thus, enzyme-loaded artificial cells
should have the advantage of not becoming involved in immunological reactions. Detailed studies on the immunological aspects
of microencapsulated enzymes show that catalase enclosed in
artificial cells is not involved in immunological reactions (Poznansky and Chang, 1969).

Patients with the inborn error of metabolism, acatalasemia,
suffer only from characteristic oral lesions due to the local action
of H_2O_2 prodced by bacterials. Experiments done in acatalasemic mice show that artificial cells containing catalase, when
applied locally to oral lesions, acted efficiently on H_2O_2 (Chang,
1972a). In the same study, artificial cells containing catalase
applied locally to the oral mucosa of an human subject (TMSC)
did not produce any irritation or adverse effects, but continued
to act efficiently on H_2O_2. With the preparation of stabilised
catalase artificial cells (Chang, 1971d), the possibility of long
term replacement therapy is even greater.

Conclusion

The results obtained indicate that in acute experiments, artificial cells containing catalase efficiently replace catalase deficiency in acatalasemic mice without being involved in immunological reactions. This model is of particular interest, for there is also a similar enzyme deficiency disease, acatalasemia, in humans.

ARTIFICIAL CELLS IN SUBSTRATE-DEPENDENT TUMOURS

Introduction

Guinea pig serum suppresses the growth of mice lymphosarcoma (Kidd, 1953) because asparaginase in the serum depletes the extracellular supply of asparagine to asparagine-dependent tumors (Broome, 1961). Asparaginase can be obtained in much larger amounts from the bacteria *Escherichia coli* for use in large-scale studies (Mashburn and Wriston, 1964). Recent reviews (Broome, 1968; Adamson and Fabro, 1968; Haskell *et al.*, 1969; Whitecar *et al.*, 1970) suggest that parenterally injected *E. coli* asparaginase is removed rapidly as foreign protein and that there appears to be a relationship between the antitumor activity and the plasma half-life of L-asparaginase; in addition, *E. coli* asparaginase may give rise to hypersensitivity and immunological reactions. Since microencapsulated enzymes can act efficiently *in vitro* and *in vivo* on permeant substrate without being involved in immunological reactions, a study has been made to examine the effects of microencapsulated asparaginase on 6C3HED lymphosarcoma-implanted mice (Chang, 1971a).

Preparation

Nylon membrane artificial cells containing asparaginase were prepared as follows: 1260 units (1 unit produces 1 μM of ammonia in 60 minutes—Broome, 1963) of the enzyme was dissolved in 1.5 ml of a 10 gm% hemoglobin (hemoglobin substrate, Worthington Co.) solution. To the 1.5 ml enzyme solution in a 100 ml beaker was added an equal volume of a freshly prepared alkaline 1,6-hexamethylenediamine solution (an aqueous solution containing 4.4 gm% 1,6-hexamethylenediamine (Eastman), 1.6 gm% sodium bicarbonate, and 6.6 gm% sodium carbonate). Immediately upon mixing the two solutions in a 100 ml beaker, the subsequent steps for the preparation of 80μ mean diameter nylon artificial cells described in this monograph were carried out. A brief outline is as follows: 15 ml of a "mixed solvent" (chloroform-cyclohexane, 1:4, containing 1% v/v of Span 85 was added to the aqueous solution. The mixture was mechanically emulsified for one minute at 4°C using a Jumbo magnetic stirrer at a speed setting of 5 to give emulsified microdroplets of 80μ mean diameters. To this stirred emulsion was added 15 ml of sebacoyl chloride solution. (Sebacoyl chloride solution 0.018M, was prepared immediately before use by adding 0.1 ml of pure sebacoyl chloride [Eastman] to 25 ml of the mixed solvent.) The reaction was allowed to continue for three minutes at the same stirring speed, then quenched by the addition of 30 ml of the mixed solvent. All the supernatant was removed by centrifugation or sedimentation, and 30 ml of a dispersing solution (equal volumes of Tween 20 and water) was added. The suspension was stirred at a speed setting of 8 for one minute. The speed was then decreased to 5, and 50 ml of water was added to the stirred suspension. The suspension was stirred for another 30 seconds and then poured into a beaker containing 200 ml of saline. After removal of the supernatant, the artificial cells were washed repeatedly in saline to remove Tween 20 and any artificial cells which were not well formed. Only properly perpared artificial cells which did not show any leakage of enzymes were used in these experiments. Control artificial cells were prepared in exactly the same way, except that asparaginase had not been added to the hemoglobin solution.

Collodion membrane artificial cells containing asparaginase were prepared as follows: 1260 units of the enzyme was dissolved in 3 ml of a 10 gm% hemoglobin (hemoglobin substrate, Worthington Co.) solution. The remaining steps for the preparation of collodion membrane artificial cells were carried out.

in vitro Studies

In vitro studies using rapid mixing and sampling technique showed that ^{14}C-labelled asparagine equilibrated rapidly across the artificial cell membranes (Chang et al., 1969). When assayed by the ammonia procedure of Broome (1963), the enzyme activity of microencapsulated asparaginase was 70 units/ml of a 50% artificial cell suspension. Thus the microencapsulated asparaginase acted on external asparagine at one-third the efficiency of the enzyme in free solution. Asparaginase, whether in free solution or in the microencapsulated form, had the same Km (Michaelis constant) values; however, the microencapsulated enzyme had a lower V_{max} value (Chang *et al.*, 1969). The microencapsulated asparaginase in aqueous suspension retained 80 percent of its original activity after three weeks of storage at 4°C.

in vivo Studies

In vivo experiments on C3HHeJ mice (Jackson Laboratories, Bar Harbour) showed that intraperitoneally injected asparaginase-loaded artificial cells lowered the blood asparagine level significantly (Chang *et al.*, 1968). These results led to further *in vivo* experiments on C3HHeJ mice (Chang, 1969e, 1971a). The procedure of tumour implantation (Broome, 1963) was modified slightly by using a tenfold amount of 6C3HED lymphosarcoma cells. Each mouse received a subcutaneous injection of 500,000 6C3HED lymphosarcoma cells (Jackson Laboratories) in each groin. Immediately after implantation, each mouse was given one of the following intraperitoneal injections: 0.05 ml/gm body weight of saline; 0.05 ml/gm body weight of a 50% suspension of control artificial cells; 0.05 ml/gm body weight of an asparaginase solution (70 units/ml solution); or 0.05 ml/gm body weight of a 50% suspension of artificial cells loaded with asparaginase (assayed activity 70 units/ml 50% suspension). The tumor implanted mice were followed as described by Broome (1963). Figure 36 shows the time when the tumor first appeared. The time when the tumor first appeared was 9.0±0.9 days (mean ± SD) for those which had received saline injections; 8.7±1.6 days for those which had received control artificial cells injections; and 14.0±4.5 days for those which had received a single intraperitoneal injection of asparaginase solution. In tumour implanted mice which had received a single intraperitoneal injection of asparaginase-loaded artificial cells, no tumours appeared in 50 percent of the implanted sites after 120 days. These results showed that when compared to the asparaginase solution, the microencapsulated form was

Artificial Cells

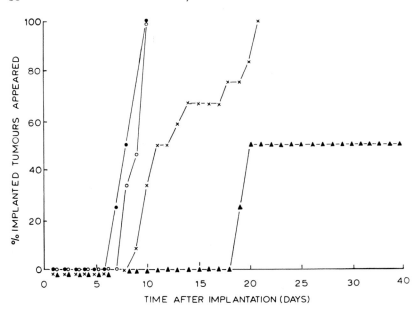

Figure 36. Appearance of implanted 6C3HED lymphosarcoma in mice.
● =saline as control, O = control artificial cells, × = asparaginase solution, ▼= artificial cells containing asparaginase.

much more effective in suppressing the growth of implanted
mouse lymphosarcoma. In order to explain this difference, one
might look at the results of experiments described in an earlier
chapter in which nylon membrane artificial cells of 80μ mean
diameter containing ^{51}Cr-labelled heterogenous hemoglobin were
injected intraperitoneally. After the first week, the artificial cells
were well dispersed over the whole peritoneal cavity. After the
second week and thereafter, there was some tendency for them
to be found in larger numbers in the upper parts of the peritoneal cavities. No significant radioactivity was detected in the
lung, liver, spleen, lymph nodes, blood, or particle free peritoneal
washings. Recovered radioactivity was associated with the artificial cells in the peritoneal cavity during the four-week follow-up.
Thus, intraperitoneally injected artificial cells of 80μ mean
diameter remained in the peritoneal cavity for at least four weeks,
and the microencapsulated hemoglobin did not leak out of the
artificial cells throughout this time. These results would seem

to support the possibility that after parenteral injection, free asparaginase was removed rapidly as foreign proteins (Adamson and Fabro, 1968; Haskell *et al.*, 1969), whereas the microencapsulated form remained inside the artificial cells in the peritoneal cavity. This way, asparaginase could continue to act on asparagine dialyzing into the artificial cells.

Artificial cells with biologically compatible membranes have been prepared (Chang *et al.*, 1968). Insolubilized enzymes like urease catalase, and trypsin have also been microencapsulated in this laboratory. This progress may eventually lead to preparations which might act indefinitely after parenteral injections. The implication of dialysis against enzymes using semipermeable artificial cells (Chang, 1966) or other modified approaches (Stanfield, 1968; Martel *et al.*, 1970) should not be confined to the specific examples mentioned. There may well be other amino acid dependent tumours, for example, serine-dependent leukemia (Regan *et al.*, 1966).

DISCUSSION

The results obtained strengthen the feasibility of using artificial cells for enzyme therapy. The ultimate clinical uses of enzyme-loaded artificial cells will depend on parallel developments in all aspects of enzyme technology. Thus, the availability of a sufficient amount of purified enzymes is the first and foremost criterion. Large-scale enzyme production, extraction, and purification from biological sources is now a well established technology. Other examples include the extraction of fibrinolysin and urokinase from homologous sources. With the recent interest in the large-scale culture of human cells, a potentially new source of homologous enzymes may be available. Other exciting progress is in the chemical synthesis of enzymes. Merrifield's group was able to adapt its automatic solid phase polypeptide synthesis technique (Merrifield, 1965) for the chemical synthesis of the enzyme ribonuclease (Gutte and Merrifield, 1969). With increasing knowledge of the basic functional and structural aspects of enzymes, this technology can be used to synthesize an increasing number of enzymes.

Whether in free solution or loaded within artificial cells, enzymes deteriorate slowly at 37°C. To have a stable system, in-solubilized enzymes (catalase, urease and trypsin) have been loaded within artificial cells (Fig. 38) (Chang, 1969d, 1971d) to produce enzyme-loaded artificial cells with prolonged activity.

The subject of immobilised enzymes has already been excellently reviewed (Silman and Katchalsky, 1966; Axen *et al.*, 1967; Mosbach 1970; Gryszkiewicz 1971), so that only a very brief summary is required here. Water-insoluble derivatives of enzymes are formed by binding enzymes to insoluble carriers by a number of approaches. Enzymes can also be physically adsorbed to inert carriers like glass beads, cellulose, or charcoal (Fig. 37). However, desorption of the enzymes may occur with changes in ionic strength, pH, temperature, or other factors. Cross-linking of the adsorbed enzymes prevents these problems (Haynes and Walsh, 1969). Another approach is to trap enzymes inside the lattice of a gel, for instance, cross-linked polyacrylamide gels (Fig. 37). In these cases, even after exhaustive washing, small amounts of enzyme still leak out. Another approach involves the covalent binding of enzymes to an insoluble carrier (Fig. 37) by a number of procedures. By this method, enzymes are attached much more firmly to their carriers. The enzymes which have been prepared in the insolubilized form include trypsin, chymotrypsin, papain, pepsin, urease, invertase, diatase, alcohol dehydrogenase, and others. Derivatives of hydrolases act on both low and high molecular weight substrates, although they show lower activity on macromolecular substrates. This has been explained as due to the steric hindrance of the carrier, which may impede the fitting of the macromolecules to the enzymes. The disparity between the action on low and high molecular weight substrates is greater with a high carrier-to-enzyme ratio: preparations with a high enzyme content do not show this disparity. Covalently bound, insolubilized enzymes like ribonuclease retain some of their ability to interact with their antibodies. One of the most attractive features of these water-insoluble derivatives is that in aqueous suspension they are much more stable than the corresponding native enzymes. For instance, insolubilized trypsin, chymotrypsin, papain, and urease retained most of their activity after storage at 4°C. for several months. Further development in the

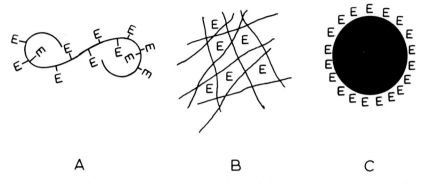

A   B   C

Figure 37. Schematic representation of insoluble enzyme derivatives. E represents an enzyme molecule. (C) Enzymes absorbed to carrier. (B) Enzymes trapped in gel lattice. (A) Enzymes covalently linked to carrier. (From Chang, 1969. Courtesy of *Science Tools*, Sweden.)

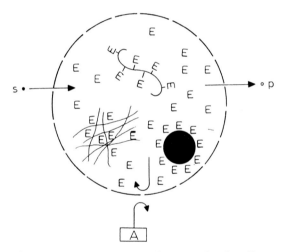

Figure 38. Schematic representation of an artificial cell containing both native enzymes and insoluble enzyme derivatives. (From Chang, 1969. Courtesy of *Science Tools*, Sweden.)

field of insolubilized enzymes will increase the scope of possible uses of enzyme-loaded artificial cells.

By loading enzymes within artificial cells, one avoids the problems associated with immunological reactions, contamination, rapid *in vivo* removal, and inactivation. In addition, when the reactions involve more than one enzyme system or where the products require removal by a nonenzyme system, one can enclose the different systems together in the artificial cells. Another advantage of enzyme-loaded artificial cells is that the sites of action can be controlled to a certain extent (Chang *et al.,* 1968; Chang, 1969d). For example, when placed in an extracorporeal shunt system, the microencapsulated enzymes can be perfused by blood, peritoneal fluid, dialysate fluid, or another body fluid; when injected parenterally they can remain at the site of final location; when administered orally, they can act in the gastrointestinal tract without being destroyed by proteolytic enzymes. This is made possible by selecting different types of biologically inert artificial cell membranes with variations in surface charge, chemical composition, and degree of blood compatibility. Some of these studies will be described in detail in later chapters.

Chapter 6

RED BLOOD CELL SUBSTITUTES

INTRODUCTION

T HE discovery of blood groups by Landsteiner and co-
workers has led to the widespread clinical use of homologous blood. The demand for blood for clinical uses is so great
that blood from donors cannot adequately fulfill this requirement. In addition, blood types are more complicated than have
been originally thought, and careful blood typing and matching
is required. There is also the possibility of transmission of infec-
tions, like infectious serum hepatitis, from donor blood. Furthermore, blood stored by the standard procedures can only be kept
for a few weeks. These and other problems have led investigators
to search for possible blood substitutes. The ideal blood substitute
should contain a fraction which can take over the function of
plasma and another fraction which can take over the function of
red blood cells. Although plasma substitutes have already been
used clinically for some years, clinically useful substitutes for red
blood cells are still not available at present. Research is being
continued to find a clinically useful substitute for red blood cells.

Oxygen in various forms has been used to substitute the oxygen-carrying functions of red blood cells. This includes the intravenous infusion of gaseous oxygen (Ziegler, 1941) or of micro-
bubbles of oxygen gas (Hori and Toyoda, 1962). Unfortunately,
infusion of oxygen leads to multiple gas emboli (Grodins *et al.,*
1943). The use of hemoglobin solutions has also been investigated. Thus Folkman's group (1966) showed that hemoglobin
solution (1 gm/100 ml) can maintain the viability of isolated
perfused organs for weeks. Others (Dean *et al.,* 1967) observed
that a patient with complete hemolysis of the circulating red
blood cell mass could survive for several hours with tissue gas
exchange occurring by means of plasma hemoglobin. Unfortunately, hemoglobin (mol wt 68,000) is rapidly cleared by the
kidneys in a matter of hours (Fairley, 1940). High concentra-

tions of plasma hemoglobin also have an adverse effect on the kidneys. Another line of approach involves the study of synthetic substitutes for the oxygen-carrying pigment hemoglobin. Wang (1958) demonstrated that several coordination compounds of heme and imidazole dissolved in organic solvents can combine reversibly with oxygen. However, in aqueous solutions, these compounds lose these properties. Minoshima (1965) investigated a water-soluble bis-histidino-cobalt compound. This compound can absorb oxygen reversibly for one hour at 37°C. Infusion of this compound into bleeded rabbits can maintain oxygen supply for about one hour. However, like natural hemoglobin in free solution, this compound (mol wt 820) is rapidly cleared by the kidney and excreted in the urine with a half-time of about 20 minutes. Thus, it is clear that in the intact animal, even if a water soluble compound with exactly the same properties as hemoglobin can be synthesized, it still cannot remain in circulation for long.

ARTIFICIAL CELLS CONTAINING ERYTHROCYTE HEMOLYSATE

Introduction

In nature, hemoglobin and complex enzyme systems are enveloped in red blood cells. The enclosing cell membranes, while impermeable to the enclosed hemoglobin and enzymes, are freely permeable to oxygen, carbon dioxide, and other small molecules. This way, hemoglobin is maintained in an intracellular environment and prevented from being excreted by the kidneys; at the same time the enclosed hemoglobin can combine reversibly with extracellular oxygen. Carbonic anhydrase, while remaining in an intracellular environment, acts on permeant extracellular carbon dioxide and thus plays an important part in the transport of carbon dioxide. Other intracellular enzymes like catalase remove peroxides which would otherwise oxidize the oxygen-carrying hemoglobin to the non-oxygen-carrying methmoglobin. Still other enzyme systems like the methemoglobin reductase systems are present in the red blood cells to convert methmoglobin to hemoglobin. This way, hemoglobin is maintained in an active state throughout the life span of the red blood cells. In

1957, I showed that the contents of red blood cells can be enveloped in ultrathin spherical artificial polymer membranes of cellular dimensions (Fig. 2). In this form, the hemoglobin and the elaborate enzyme systems normally present in the red blood cells are now enveloped in an "intracellular environment." This way, hemoglobin and enzymes are prevented from leaking out to be excreted or to become involved in immunological reactions. At the same time, oxygen and carbon dioxide can equilibrate rapidly across the membrane to interact with the enclosed materials. If these artificial cells are nontoxic and capable of circulating for a long time in the bloodstream, the general principle of artificial cells might be applicable for red blood cell substitutes, since in this form, the membranes of the artificial red blood cells would not become fragile after prolonged *in vitro* storage in the blood bank. The artificial red blood cells could be transfused without cross-matching, since they would lack the ABO and other blood group antigens that are attached to the red blood cell membranes. Even though this approach is far from being ready for clinical use, some progress has been made.

Hemoglobin And Enzymes

The *in vitro* biochemical properties of artificial cells containing red blood cell hemolysate have been studied. Earlier studies have shown that red cell hemolystate enveloped in collodion membrane artificial cells contains a combination of oxyhemoglobin and methemoglobin and that the oxyhemoglobin retains its ability to combine reversibly with oxygen (Chang, 1957, 1964). More recent studies (Chang and Poznansky, 1968a) show that if additional catalase has been enclosed along with the red blood cell hemolysate, there is a marked reduction in methemoglobin formation. In the case of nylon membrane artificial cells, the preparative procedure results in the formation of a much higher proportion of methemoglobin (Chang, 1964, 1965). Red cell hemolysate enclosed in Silastic rubber artificial cells (Chang, 1966) remains unaltered and can combine reversibly with oxygen, and it retains this ability for at least a few days of storage at 4°C. More recently, studies of a Japanese group have also supported these findings. Hemoglobin microencapsulated within polymer membranes like polystyrene, ethyl cellulose, and dextran stearate (Sekiguchi and Kondo, 1966) has been shown to combine reversibly with oxygen (Toyoda, 1966).

Carbonic anhydrase, present in high concentration in red blood cells, plays an important part in the transport of carbon dioxide. The carbonic anhydrase activity of erythrocyte hemolysate enclosed in artificial cells of collodion, nylon, or cross-linked protein membranes has been analyzed (Fig. 39) (Chang, 1964, 1965). The carbonic anhydrase activity was measured as the rate of fall of pH when carbon dioxide was bubbled through a buffered suspension (Philpot and Philpot, 1936). The rate of fall of pH was accelerated by the presence of erythrocytes, or of artificial cells containing red blood cell hemolysate. Acetazolamide (Diamox®) in the concentration used, completely inhibited the catalytic activity of both the erythrocytes and the nylon, cross-linked protein, or collodion membrane artificial cells. The samples of buffered saline that had been in contact with their own volume of artificial cells for 24 hours showed no detectable enzyme activity, indicating that the enclosed carbonic anhydrase had not leaked out during this period. Carbonic anhydrase in artificial cells had about 75 percent of the activity of the erythrocyte-bound enzyme. Thus, carbonic anhydrase in artificial cells acts efficiently in catalyzing carbon dioxide.

Catalase is another important component of mammalian erythrocytes. Its enzymatic activities in the hemolysate enclosed in collodion membrane artificial cells has also been analysed (Chang, 1967; Chang and Poznansky, 1968a). Both its natural substrate, hydrogen peroxide, and its chemical substrate, sodium perborate, have been used. Catalase enclosed in collodion membrane artificial cells acts efficiently *in vitro;* it also acts efficiently *in vivo* in replacing the enzyme in mice which have a congenital deficiency of catalase in their red blood cells (Chang and Poznansky, 1968a).

Immunological Studies

Human ABO blood group antisera cause marked agglutination of both incompatible human erythrocytes and heterogeneous erythrocytes. On the other hand, studies in this laboratory show

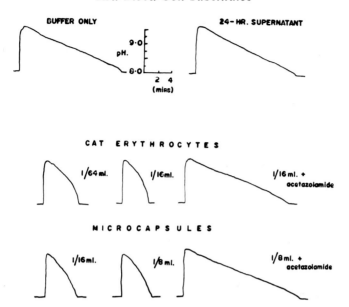

Figure 39. Carbonic anhydrase activity, measured as the rate of fall of pH in the presence of CO_2. Fall of pH is accelerated by erythrocyte or artificial cells containing hemolysate. Acetazolamide eliminates the enzyme activity in erythrocyte and artificial cells. Supernatant from artificial cells has no enzyme activity. (From Chang, 1964. Copyright 1964 by the American Association for the Advancement of Science.)

that these antisera do not cause any agglutination of artificial cells containing red blood cell hemolysate, even though the hemolysate has been obtained from incompatible human erythrocytes or heterogeneous erythrocytes. Thus artificial cells do not appear to possess the blood group anigens normally present in biological cell membranes.

Intravenous Injection

The fate of intravenously injected nylon membrane artificial cells containing hemolysate has been studied (Chang, 1964, 1965; Chang and MacIntosh, 1964a). A suspension of $5\mu \pm 2.5\mu$ diameter (mean ± standard deviation) nylon membrane artificial cells was infused intravenously over a period of 1.5 minutes, and arterial samples were obtained from the anesthetized cat during and following the infusion. The arterial level rose steeply during infusion, but upon discontinuing the infusion, the level fell (Fig. 40) at first with a half-time of about 1.6 minutes and thereafter somewhat more slowly. This result made it obvious that though some of the artificial cells successfully completed one or more circulations, they were removed rapidly from the bloodstream. Experiments using ^{51}Cr-labelled nylon membrane artificial cells show that most of them are removed by the lung, liver, or spleen. Collodion membrane artificial cells are also removed rapidly from the circulation. Thus it appears that nylon and collodion membrane artificial cells of 5µ average diameters are rapidly removed from the circulation. During the infusion there is a fall in arterial blood pressure and an increase in venous pressure and other signs compatible with pulmonary embolism.

At this point, it is interesting to note that the formed elements of blood of comparable dimensions survive for a long time in the circulation. From the experiments of Harris (1963), Rand and Burton (1963), Canham and Burton (1968), Jandl *et al.* (1961), and others, there is no doubt that the deformability of red blood cells is an extremely important factor in allowing them to pass through the capillaries of smaller diameters. In addition, very minute changes in the shape of erythrocytes also affect their survival in the circulation (Canham and Burton, 1968). Immunological properties is another obvious factor in the survival of blood cells in the circulation. Immunologically incompatible red blood cells are removed rapidly from the circulation by the reticuloendothelial system (Halpern *et al.*, 1958). However, in the case of artificial cells, the membranes are devoid of antigens, and *in vitro* experiments in this laboratory described earlier show that they are not agglutinated by blood group sera.

Although the size of the particles is important (Ring *et al.*, 1961), size alone is not the only determining factor, since foreign particles as small as 25 Å are removed rapidly from the circulation by the reticulo-endothelial system (Halpern *et al.*, 1958). Another factor that might be important is the effect of surface properties of artificial cells (Chang, 1964, 1965; Chang and MacIntosh, 1964a). Red blood cells and other formed elements of blood have a negative surface charge (Abramson, 1934) which

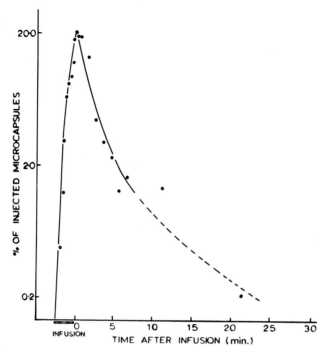

Figure 40. Survival of nylon membrane artificial cells (mean diameter 5μ) after intravenous infusion into cats; arterial sampling. (From Chang, 1965.)

is due to the presence of n-acetylneuraminic acid containing mucopolysaccharide on the red blood cell membrane (Cook *et al.*, 1961). Platelets are known to contain a sulfated polysaccharide. Rambourg *et al.* (1965) have demonstrated the presence of a well-marked carbohydrate coating on many mammalian cells. Davies (1958, 1961) reported that the electrophoretic mobility of erythrocytes was diminished in myocardial infarction, and suggested that such a reduction of surface charge would favour their adhesion to one another or to blood vessel walls. Danon and Marikovsky (1961) found that the younger erythrocytes were about 30 percent more strongly charged than the older ones obtained from the same blood sample. All this information leads us to contemplate the importance of the physicochemical surface properties in allowing a particle to survive in circulation (Chang, 1965). Investigation was carried out in this laboratory to study the relationship of surface properties to the survival of erythrocytes, foreign particles, and artificial cells in circulation.

One of the experiments involved the study of the effects of removal of neuraminic acid from red blood cells on their survival in circulation. Experiments on the survival of ^{51}Cr-labelled erythrocytes were carried out in four dogs according to the plan shown in Figure 41. The animals' own erythrocytes were injected on three occasions. On the second occasion, the reinjected erythrocytes had been treated with neuraminidase to remove neuraminic acid by the procedure of Cook *et al.* (1961), washed with saline, tagged with ^{51}Cr according to standard procedures (Chien and Gregerson, 1962), and then washed three times with saline.

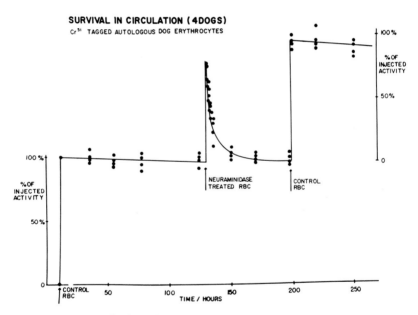

Figure 41. Survival of reinfused autologous dog erythrocytes, tagged with [51]Cr. Data from four dogs; each point represents the radioactivity of one blood sample from a limb vein. In each case the first and third tests were with normal erythrocytes and the second test was with neuraminidase-treated erythrocytes. Ordinates give percent of injected label in circulation at time of sampling, asuming that blood volume, calculated as the dilution volume of the injected label in the first test, was the same throughout. Note rapid diasappearance of neuraminidase-treated erythrocytes. (From Chang, 1965.)

On the first and last occasions, the erythrocytes were treated similarly except that no neuraminidase was added. The results are shown in Figure 41. The control cells disappeared very slowly from the circulation, compatible with the usually stated lifetime of about three to four months for red cells of dogs (Berlin *et al.*, 1959). The neuraminidase-treated erythrocytes fell exponentially with a half-time of only about two hours. The response of each animal to the final injection of control erythrocytes was not significantly different from its response to the first injection. These results led to our suggestion that the survival of erythrocytes may be related to the neuraminic acid present in the membranes; thus older cells with a decrease in neuraminic acid are removed from the circulation (Chang, 1965; Chang and MacIntosh, 1964).

These results led to the investigation of a simpler model system (Chang, 1965). Polystyrene latex particles (Dow Chemical Co.) $2.05\mu \pm 0.018\mu$ (mean $\pm$ standard deviation) were diluted 1:100 in saline. A similar suspension was prepared using polystyrene latex sulfonated for 30 seconds by Ag_2SO_4 and H_2SO_4, and then washed with saline (Chang, 1965). The arterial blood levels of the control and the sulfonated polystyrene latex after intravenous injection are shown in Figure 42. With the sulfonated polystyrene latex, the initial counts were close to the expected number; after this the level declined, with a half-time of about 1.6 minutes, and then apparently more slowly. In the case of the control polystyrene latex particles, even in the first sample collected, half a minute after the injection the arterial blood contained only 10 to 15 percent of the theoretical number of particles, and then fell with a much steeper slope with a half-time of about 15 to 30 seconds. Further experiments show that the sulfonated polystyrene latex particles are not trapped to any extent by the lung, but are removed by the extrapulmonary tissues like the portal circulation. Control polystyrene latex particles, on the other hand, are trapped fairly efficiently by the lung, and very efficiently by the extrapulmonary tissues, especially the portal circulation. Wilkins and Meyers (1966) have also reported similar studies. In their experiment, the surface charge of polystyrene particles was altered by the adsorption of macromolecules: gum arabic for negative charge and polylysyl gelatin for positive charge.

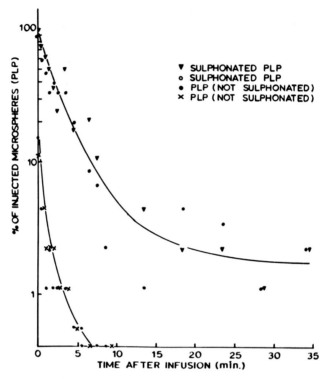

Figure 42. Survival of polystyrene microspheres infused into cats. (Infusion took 30 seconds ending at zero time.) Data from four experiments, two with untreated and two with sulfonated microspheres. Ordinates give arterial counts of artificial cells, as percentage of the count expected on the assumption that all injected artificial cells would be uniformly distributed in a blood volume of 70 ml/kg. (From Chang, 1965.)

They demonstrated a relationship between surface charge and site of removal. Negatively charged polystyrene particles are largely taken up by the liver, while positively charged polystyrene particles are initially accumulated appreciably in the lungs.

Artificial cells were prepared with sulfonated nylon membranes (Chang, 1965) for the study of the effects of charge on surface properties. Sulfonated nylon membrane artificial cells prepared as described had a negative surface charge comparable to that of erythrocytes. Preliminary results showed that although sulfonated nylon artificial cells are also removed rapidly, they sur-

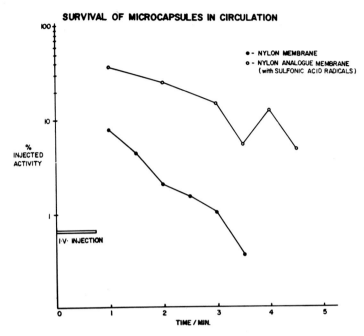

Figure 43. Survival of labelled nylon membrane artificial cells (mean diameter 5μ) after intravenous infusion into cats; arterial sampling. Upper line: sulfonated nylon membrane artificial cells. Lower line: ordinary nylon membrane artificial cells. Survival estimated as percentage of radioactivity expected on the basis of uniform distribution of the label in a blood volume of 70 ml/kg. (From Chang, 1965.)

vive significantly longer than unsulfonated nylon membrane artificial cells (Fig. 43). In addition, the unsulfonated nylon membrane artificial cells are less liable to be trapped in the pulmonary circulation and the reticuloendothelial system than the control ones.

Since natural erythrocytes and the endothelium of vessels are covered by mucopolysaccharide, further studies were done in this laboratory to incorporate a complex containing a mucopolysaccharide (heparin) into the artificial cell membrane (Chang *et al.*, 1967b) and to study their survival in the circulation. Collodion membrane artificial cells prepared by the standard procedure, and heparin-complexed collodion membrane artificial cells were used. Preliminary results showed that while both types of artificial cells were removed rapidly from the circulation, those with a heparin-complexed membrane survived significantly longer.

The significant difference in length of survival of sulfonated polystyrene latex, sulfonated nylon membrane artificial cells, and heparin-complexed artificial cells, when compared to their respective control counterparts, seems to indicate that surface properties do play an important part. It seems that surface negative charge per se may not be the deciding factor. The effective surface charge of a suspended particle depends very much on the nature of the fluid in which it is suspended. For instance, Seaman and Swank (1963) observed that althouhg lipid emulsions of varying pH had electrophoretic mobilities in saline that corresponded to their pH, the same lipids emulsified in plasma had identical mobilities. Likewise, Mirkovitch *et al.* (1964) found that the negative surface charge of sulfonated polystyrene was lost almost immediately when suspended in dilute plasma. The type of surface may decide the selective adsorption of certain types of plasma components onto its surface, thereby deciding whether the particles are suitable for prolonged survival in the circulation.

SILICONES AND FLUOROCARBONS

Another line of approach involves the use of a class of inert organic chemicals called silicones and fluorocarbons. Silicones are organic compounds with organic groups added to a skeleton chain of silicone organic groups. The silicones are available in the form of fluids, resins, and rubbers. Flourocarbons are organic compounds in which the hydrogen atoms have been replaced by fluorine. Both groups have a high solubility for oxygen and other gases.

Polymerized Emulsions Of Silicones

A type of silicone can be formed by adding a catalyst (e.g. stannous octoate) to a thick liquid polymer, Silastic RTV. The resulting silicone is in the form of a rubber that has useful mechanical properties and is chemically inert (Braley, 1960). Like other types of silicones, it is highly soluble to oxygen and carbon dioxide. Experiments have been done to test this group of materials for use as artificial red blood cells (Chang, 1966). This is done by forming a double emulsion of hemolysate and polymethylsiloxane to form microspheres, each consisting of 1 part of hemolysate and 2 parts of silicone rubber. In this emulsified form both the silicone rubber and the enclosed hemoglobin can combine reversibly with oxygen; in addition, the presence of carbonic anhydrase in the silicone rubber would make the carriage of carbon dioxide much more efficient. Also, in this emulsified form, the artificial cells can be suspended in an aqueous solution which can carry the water soluble nutrients like glucose and amino acids required for the metabolism of tissue cells. When infused intravenously into cats, the silicone rubber artificial cells (mean diameter 4μ) were removed rapidly from circulation. A number of factors responsible for their rapid removal may include their rigidity and size.

Silicone Oils and Fluorocarbon Oils

Clark and Gollan (1966) immersed animals in oxygen-saturated silicone oils. When immersed in silicone oils having a viscosity of 1 centistoke, mice survived for 35 minutes at 24°C and six hours at 18°C; cats survived for one hour. When animals were immersed in oxygen-saturated fluorocarbon liquid FX-80, their arterial oxygen tensions were maintained between 140 and 300 mm; their arterial pCO_2 increased from 50 to 80 mm and their arterial pH fell from 7.35 to 7.10. Thus although the arterial oxygenation was good, there was some impairment of carbon dioxide elimination. Immersion in the fluorocarbon liquid resulted in pulmonary damages. Gollan and Clark (1966) perfused isolated rat hearts alternately with oxygenated FX-80 fluorocarbon and oxygenated diluted blood. They found that the isolated rat hearts continued to beat strongly. Work in the same laboratory (Linn *et al.*, 1967) involved the use of silicone oils (1 to 50 centistokes viscosity) or fluorocarbon oils to perfuse isolated kidneys. Perfusion with either of these liquids allowed the isolated kidney to be stored for a significantly longer period of time. However, the organic liquid can only carry oxygen and carbon dioxide, and not water-soluble materials and metabolites such as glucose, amino acids, vitamins, and electrolyes.

Emulsion Of Fluorocarbon Oils

In 1967, instead of using microscopic particles of silicones or organic fluid, Sloviter, Clark, Geyer, and other groups investigated the use of microscopic emulsion of fluorocarbon oils suspended in aqueous solutions. Thus Sloviter's group (Sloviter and Kaminoto, 1967; Andjus *et al.*, 1967) used ultrasonic treatment to emulsify FX-80 in a Kreb-Ringer buffer solution containing albumin. Unlike artificial cells made from silicone rubber, the emulsified fluorocarbons were smaller in diameter (2μ diameter) and nonrigid. Sloviter and Kaminoto (1967) used this suspension to perfuse isolated rat brain preparations. A 20% suspension of either erythrocytes or fluorocarbon emulsions was suspended in bovine albumin solution containing 200 mg/100 ml glucose. Experiments were done at 25°C using 5% carbon dioxide and 95% oxygen. When the isolated brain was perfused with the aqueous solution alone, spontaneous EEG ceased after five minutes. When blood cells or fluorocarbon emulsion was added, the spontaneous EEG persisted for two hours or more. The arteriovenous difference in oxygen tension was 300 mm for blood cells and 200 to 250 mm for FX-80 emulsions. The arteriovenus differences in pH and carbon dioxide tension were the same for blood cells and FX-80 emulsions. These results demonstrate that the FX-80 emulsion performs efficiently in the isolated brain preparation and suggest the exciting possibility that these preparations might be used for perfusion of other organs. Thus isolated rat livers have been perfused with emulsions of fluorocarbon FX-80 and FC-43 (Triner *et al.*, 1970) for periods of two hours. There were no signs of deterioration reflected in the parameters tested.

In earlier studies by Sloviter's group it was found that when these preparations were infused intravenously into intact animals like rats, cats, hamsters, rabbits, chickens, and dogs, the animals died within a matter of hours. Autopsy findings of these animals showed distended lungs with hemorrhagic areas and dilated right hearts. On the other hand, Clark (1967) showed that one-third of the blood of a rabbit could be replaced by FC-43 emulsion in Pluronic® F68 and rabbit plasma.

Using ultrasonic dispersion and nonionic detergents, Geyer *et al.* (1968) prepared finer emulsions of 1μ and less in diameter. They found that the finer FX-80 emulsions were less acutely toxic, although intravenous injections still caused the death of rats in a matter of hours. Autopsy findings showed that the lungs were grossly distended. Geyer's group then tested another fluorocarbon, FC-43, which was similarly dispersed into fine emulsion. This FC-43 emulsion contains 37 vol% oxygen at a pO_2 of 711. Unlike red blood cells, the dissociation curve of FC-43 emulsion is linear. This means that in an atmosphere of 100% oxygen, an aqueous suspension of FC-43 has an oxygen-carrying capacity comparable to that of red blood cells, but at atmospheric partial pressures, the FC-43 suspension is less efficient. Having demonstrated that this type of fluorocarbon FC-43 emulsion had no toxic effects after repeated intravenous injections, Geyer's group proceded further. They used a preparation containing emulsions of FC-43 in high molecular weight nonionic detergent Pluronics. This preparation was used to exchange with all the red blood cells of the rats. When such animals were kept in a 100% oxygen atmosphere, their respiratory rate remained normal. In addition, these animals remained alert and active, being able to urinate, defecate, wash, and run around. This clearly demonstrates the efficiency of oxygen carriage by FC-43. These animals survived for up to eight hours in 100% oxygen. After this, the respiratory rates rapidly decreased and finally stopped. This exchange with fluorocarbon FC-43 and Pluronics did not produce any edema or accumulations of fluid in the pleural or peritoneal cavities. There were no obvious pathological findings in the organs. Furthermore, the fluorocarbon FC-43 and the nonionic detergents did not affect the metabolism of tissues, since the oxygen uptake of slices of heart, liver, and kidney was unaffected by the presence of emulsified fluorocarbon in the medium.

Further work by Sloviter's group (Sloviter *et al.*, 1969) demonstrated that unlike other animals, frogs could receive intravenous injections of FX-80 emulsions without adverse effects. Mice receiving FX-80 also survived for a number of days, although gross pathology of lungs and hearts were observed in some cases. They infused emulsions of FX-80 or FC-43 into the circulation of these animals and followed their fates in the circulation. In the case of frogs, an average of 66 percent of the injected FX-80 remained in circulation after three days. In the case of mice, about 50 percent of the injected FX-80 and about 25 percent of the injected FC-43 remained in circulation after the same length of time. Six days after injection, all of the fluorochemicals disappeared from the circulation of mice; on the other hand, a large amount still remained in the frogs. Further experiments were carried out using carbon monoxide to convert about 90 percent of the animal's hemoglobin to carbon monoxyhemoglobin. Under these conditions frogs and mice given fluorochemicals survived for significantly longer periods of time when compared to untreated controls. They (Sloviter *et al.*, 1970) continued further studies to analyse the reasons for the toxic effects of FX-80 in animals other than frogs and mice. Marked right heart dilatation and increased pulmonary resistance to flow was observed even when FX-80 emulsions of less than 2μ diameter were injected intravenously into rats. This could not be eliminated by prior treatment with antihistamine, serotonin blocking agents, isoproterenol, or large doses of heparin. Following intravenous injection of FX-80 there was a marked decrease (greater than 50% decrease) in systemic platelet levels, without any significant changes in fibrinogen or leucocyte levels. The authors suggested that multiple platelet microemboli reaching the lungs might be responsible for the observed adverse effects of FX-80 injection.

Nose's group (1970) studied in great detail the effects of intravascular fluorocarbon liquids, including emulsions of fluorocarbon. Intra-aortic injections of fluorocarbon FC-43 emulsions into unanesthetized dogs resulted in grand mal seizures. The coma lasted 20 minutes with 2μ emulsion and 30 minutes with 5μ emulsion. Incontenence, salivation, decrease in blood pressure, and increase in respiratory rates were observed. When the fluorocarbon emulsions were given intravenously to two dogs (1 ml/kg), decrease in blood pressure, increase in respiratory rate, vomiting, and incontinence were observed. There was no convulsion, but the dogs remained lethargic for 10 to 20 minutes. Laboratory tests showed an increase in LDH, SGOT and white blood cells six hours after infusion. Microcirculation studies showed that injection of 5μ

emulsion resulted in mechanical obstruction of capillary circulation. Injection of 2μ emulsion caused a temporary reduction of capillary flow lasting for 15 to 20 minutes. Furthermore, microscopic evidence of intravascular clotting was observed.

Recently, Clark *et al.* (1970) studied the possible uses of fluorocarbon emulsion less than 1μ diameter as red blood substitutes for perfusion of whole animals, using the Clark bubbledefoam heart-lung machine. A very high venous pO_2 of 100 or more was obtained. There was some acidosis, bradycardia, and ventricular arrhythmia during infusion of the fluorocarbon emulsions. During perfusion, the cerebral cathode current increased threefold. These results indicate the possible uses of fluorocarbon to prime heart-lung machines, and at the end of the procedure the emulsion can be replaced by the patient's blood.

Improvement in the perfusion mixtures were investigated by Geyer's group (1970). They prepared an extremely fine fluorocarbon FC-43 dispersion with particle diameters in the colloidal range. They replaced the blood of rats by this fluorocarbon-Pluronics mixture until the red blood cells of the animals were lowered to a hematocrit of 6 percent. If the animals were kept in a 100% oxygen atmosphere for 48 hours, the oxygen tension could be gradually reduced over a four-day period, until finally room air was used. The animals not only survived this, but the hematocrit could return to the normal values and the animals continued to grow. This clearly demonstrates the efficiency of the fluorocarbon-Pluronics mixture as an artificial blood substitute. Further lowering of hematocrit to less than 6 percent was not compatible with survival (Fig. 44). However, when serum was present in the fluorocarbon-Pluronics mixture in an amount of about 20% to 30% volume, the hematocrit could be lowered to 3% to 4%, with the rats surviving and regenerating their plasma proteins and red cells (Fig. 45). This is the most exciting demonstration of the possible uses of fluorocarbon for red blood cell substitute.

DISCUSSION

At present, the most immediate possibilities for a red blood cell substitute appear to be some forms of organic compounds which

Artificial Cells

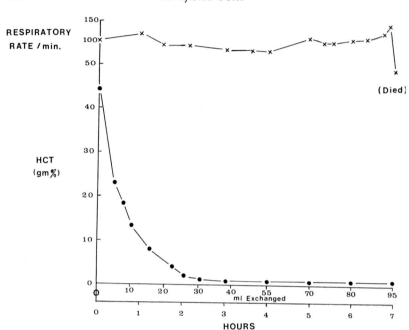

Figure 44. Blood replacement with fluorocarbon FC-43-Pluronic colloid with no serum added. (From Geyer, 1970. Courtesy of the American Federation of Biological Sciences.)

can carry oxygen and carbon dioxide. It has been pointed out (Sloviter, 1970) that these chemicals may be useful in conditions with low oxygen tension or extremes of temperature, or in situations involving hemolytic agents or agents which convert hemoglobin to an inactive form. On the other hand, the same author pointed out, the fluorochemical cannot provide the buffering and the enzymatic activity as in the case of erythrocytes. The most immediate possibilities are the perfusion of isolated organs or whole body perfusion. In addition to the oxygen-carrying capacity of the fluorocarbon, the Pluronics, as used by Geyer's group (1968), provide effective colloid osmotic pressure for the mixture. In organ preservation, the perfusing fluorocarbon-Pluronics mixture can be removed and replaced by blood before transplantation. On the other hand, when infused into intact animals as in transfusion, fluorochemicals are not metabolized by the

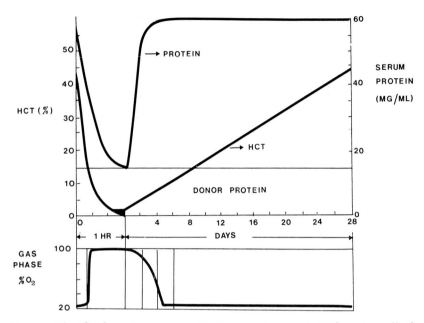

Figure 45. Blood replacement with fluorocarbon FC-43-Pluronic colloid plus 25% serum. The rat was exchange transfused via the exterior jugular vein with the fluorocarbon FC-43-Pluronic preparation which contained 30% fresh rat serum. When the hematocrit reached 3 or 4 the exchange was stopped, the catheter was sealed, and the animal was allowed to recover as described in the text. The gas phase furnished the animal is shown in the bottom portion of the figure. The dark area at the bottom of the hematocrit curve emphasizes the relatively small number of cells which remain at the end of the exhcange procedure. (From Geyer, 1970. Courtesy of the American Federation of Biological Sciences.)

body. Their fates in the body are still not known. For instance, what is the effect of a large amount of fluorocarbon accumulating in the reticuloendothelial system? A natural vegetable oil like safflower oil which can be metabolized has been suggested (Hutson *et al.*, 1968).

Artificial cells containing the entire contents of erythrocytes from a homologous source may solve the problems mentioned, since they can be metabolized by the body; they possess the buffering and enzymatic activity of red blood cells. Artificial cells containing the contents of erythrocytes have already been prepared (Chang, 1957, 1964). *In vitro,* the enclosed hemoglobin can combine reversibly with oxygen, and carbonic anhydrase can act efficiently on carbon dioxide. The enclosed catalase can act efficiently *in vivo* to replace the deficient catalase in red blood cells (Chang and Poznansky, 1968a). Before they can be used to substitute red blood cells, one has to solve the problem of producing a preparation which can survive for a sufficient length of time in circulation. More serious biomedical problems than this have been solved by concentrated efforts.

Chapter 7

ARTIFICIAL CELLS PERFUSED BY BODY FLUIDS

INTRODUCTION

EXCEPT for blood cells, which are but a minute portion of cells in the body, the majority of cells remain stationary within tissues or organs. While retained within tissues or organs, cells are perfused by blood and body fluids. A somewhat similar artificial system has been proposed (Chang, 1966). Here artificial cells are retained within a shunt chamber by screens placed on either side (Fig. 46). Blood (Chang, 1966), peritoneal fluid (Chang and Poznansky, 1968a), dialysate (Chang et al., 1968; Sparks et al., 1969), and other body fluids (Chang et al., 1968) can freely recirculate through the shunt chamber to come in direct contact with the artificial cells. In this way (1) the rate of action of the material enclosed inside the artificial cells is limited only by the rate of circulation and the diffusion of substrates across the artificial cell membranes; (2) artificial cells do not leave the shunt chamber to disperse in the body; (3) artificial cells can be renewed when required without accumulation in the body.

TYPES OF BODY FLUIDS FOR PERFUSION

Perfusion By Blood

In the earlier chapters it has been demonstrated that artificial cells act quite efficiently when injected intraperitoneally into the experimental animals. In certain cases it might be desirable to have the artificial cells in direct contact with blood-borne substrates, so that their actions are not limited by the rate at which substrates diffuse into and within the peritoneal cavity. The intravenous injection of such artificial cells will place them in direct contact with blood borne substrates. However, intravenously injected foreign particles may embolise in the pulmonary vessels or become trapped by the reticuloendothelial system.

113

Furthermore, with repeated injections, a large bulk of artificial cells may accumulate in the body.

A typical example of how these problems can be overcome has been described (Chang, 1966). Artificial cells containing urease (larger than 90μ in diameter) are retained in the shunt chamber by mesh screens placed at either end of the chamber (Fig. 46). The inlet of each shunt is connected to the femoral artery of a heparinized dog. Blood entering the shunt through the inlet screen comes into direct contact with the artificial cells. Blood urea diffusing into the artificial cells is converted into ammonia by the enclosed urease. Blood than returns to the femoral vein of the animal through the exit screen. As will be described later, this system acts very efficiently in converting substrates to products—at least ten times faster than the intraperitoneal route. Artificial cells containing ion exchange resin (Chang, 1966) and artificial cells containing activated charcoal (Chang, 1966, 1969e; Chang et al., 1970) have been similarly used.

Recirculation Of Peritoneal Fluid

Another way artificial cells can be perfused by body fluid is that in which the peritoneal fluid is recirculated continuously through an extracorporeal shunt system containing artificial cells (Chang et at., 1968; Chang and Poznansky, 1968a). A typical experiment is one in which anesthetized acatalasemic mice were each given an intraperitoneal injection of 4.5 ml of Inperinol® fluid (Abbott). This intraperitoneal fluid was continuously recirculated through a smaller (6 ml), modified version of the extracorporeal shunt chamber (Chang, 1966) for 20 minutes after the subcutaneous injection of perborate. In one group of animals the extracorporeal shunt chamber contained 4 ml of a 50% suspension of artificial cells containing catalase, and in the control group no artificial cells were present in the shunt. The results obtained showed that this system acted quite efficiently on body substrates. Thus in the control group of animals with catalase deficiency, 76.1% ± 13.4% SD of the injected substrates could be recovered after 20 minutes. When the chambers contained microencapsulated catalase, only 32.0% ± 9.8% of the subcutaneously injected substrates remained in the catalase deficient animal after

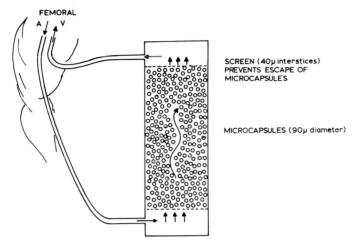

FEMORAL
A   V

SCREEN (40µ interstices)
PREVENTS ESCAPE OF
MICROCAPSULES

MICROCAPSULES (90µ diameter)

Figure 46. Artificial cells in an extracorporeal shunt chamber perfused by recirculating blood from animal. (From Chang, 1966.)

the same length of time. No catalase activity could be detected in the intraperitoneal fluid which had been recirculating through the catalase-loaded artificial cells' shunt chamber for 20 minutes, showing that no detectable enzyme had leaked out of the artificial cells throughout the experiment. Although this system is far less efficient than perfusion by blood, heparinization and arteriovenous shunts are not required.

Recirculation Of Other Body Fluids

It is obvious that dialysate fluid, lymph, and other body fluids can be similarly used to perfuse the artificial cells. For instance, artificial cells have been placed in the dialysate compartment of the artificial kidney (Chang *et al.*, 1968; Sparks *et al.*, 1969). In this way, uremic metabolites crossing the dialysis membranes can be acted on by the artificial cells.

NONTHROMBOGENIC ARTIFICIAL CELLS

The use of artificial cells in an extracorporeal shunt chamber avoids the *in vivo* introduction and accumulation of artificial cells and permits a more rapid exchange with blood-borne substrates. On the other hand, this system requires whole body or regional heparinization with the attendant inconveniences and risks of hemorrhage or toxicity of the heparin antagonist. For this reason, we have investigated the possibility of eliminating the need for a soluble anticoagulant in the extracorporeal shunt procedure.

Recent interests in the fields of biomaterials have been concentrated on the development of blood-compatible surfaces. At present, different systems are being investigated. One approach involves the binding of heparin to surfaces. Following the discovery that the binding of heparin to a surface can make it nonthrombogenic (Gott et al., 1964), many workers have developed and extended this principle. Studies of the effects of surface charge on blood clotting (Sawyer and Pate, 1953; Sharp et al., 1965) have led to the development of another class of surfaces with ionic radicals or imposed electrical charges. A third approach involves the study of materials with inert surfaces like silicone rubber, collagen (Rubin et al., 1968), albumin (Lyman et al., 1970), and hydrogels. A large amount of excellent work along these lines has been reported; unfortunately, it is not within the scope of this monograph to review these studies. We have reported the use of all three approaches for the preparation of artificial cell membranes: heparin binding, surface charge, and inert material like silicone rubber or albumin coating.

Preparation Of Artificial Cells With Heparin Binding

The shunt chamber and its attached tubing can be made nonthrombogenic by the procedure devised by Gott and his colleages (Gott et al., 1964), in which a complex of heparin with graphite and benzalkonium is applied to the polymer surface. However, this treatment and others (Leininger et al., 1966; Fourt et al., 1966; Merrill et al., 1966) embodying the same principle cannot be applied to the artificial cell surface because the enclosed enzyme or detoxicant would be inactivated. Procedures have been developed for enclosing enzymes or detoxicants in selectively permeable nonthrombogenic artificial cells (Fig. 47) Chang et al., 1967).

A benzalkonium-heparin complex (BHC) was prepared as described by Fourt et al. (1966) by the dropwise addition, with stirring, of 20 ml of a 10% aqueous benzalkonium chloride solu-

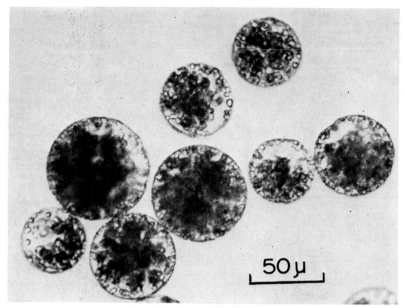

Figure 47. Artificial cells with heparin-complexed membranes. (From Chang *et al.*, 1967. Courtesy of the National Research Council of Canada.)

tion (made by diluting benzalkonium chloride solution, U.S.P. 21.3%, with water) to an equal volume of an aqueous solution of heparin sodium (U.S.P. 8 mg/ml). We found that the following additional steps were necessary to remove traces of free benzalkonium, which causes lysis of erythrocytes and leucocytes. The BHC precipitate was washed five times with 8 volumes of distilled water, then stirred for one hour with 8 volumes of water, allowed to sediment, resuspended in 3 volumes of heparin sodium solution (8 mg/ml), and stirred again for one hour. After two further washings with 3 volumes of water, the BHC was dried in a desiccator at room temperature for at least 24 hours. A 150 mg portion of it was then dissolved, with vigorous mechanical stirring, for at least three hours in 25 ml of an ether-collodion solution prepared by removing most of the alcohol from collodion (U.S.P.) and redissolving it in ether as previously described (Chang, 1965; Chang *et al.*, 1966). (It is important at this stage to continue the stirring until visible particles of BHC have disappeared and the ether-collodion solution appears quite clear. If the liquid remains turbid, the reason is most likely due to inadequate drying of the benzalkonium-heparin complex.) This BHC-collodion liquid was then substituted for the collodion solution used in our standard procedure (Chang, 1965; Chang *et al.*, 1966) for preparing collodion membrane artificial cells. As in that procedure, the aqueous protein solution to be encapsulated was mechanically emulsified in water-saturated ether; the membranes were then formed by the addition of BHC-collodion liquid to the stirred solution. For the preparation of relatively large BHC-collodion membrane artificial cells (mean diameter about 100μ) for inclusion in an extracorporeal shunt, the emulsifying agent, Span 85, was omitted and the jumbo stirrer was set at speed 2; smaller BHC-collodion membrane artificial cells (mean diameter about 5μ) for electrophoretic studies were prepared with Span 85 and a Virtis homogenizer. The use of a larger proportion of BHC results in the production of membranes that tend to lyse erythrocytes, and a smaller proportion of BHC yields membranes that are incompletely nonthrombogenic. The artificial cells finally harvested in aqueous suspension must be washed repeatedly in the centrifuge to remove traces of the Tween 20 detergent, which is strongly hemolytic; absence of all foaming when the supernatant is shaken is a satisfactory test.

Material to be microencapsulated together with the erythrocyte hemolysate can be added to the hemolysate before emulsification. For example, urease may be encapsulated by dissolving in 5 ml of unbuffered hemolysate 900 mg of urease (Sigma type III, 1720 units/gm) and 300 mg of tris buffer. This solution is filtered, and 3 ml of the filtrate is used in the procedure described above. In the case of activated charcoal, 100 gm (British Drug Houses) is suspended in the 3 ml of hemolysate before microencapsulation.

Artificial cells with a nonthrombogenic surface are also made by coating preformed artificial cells with BHC-collodion, or by trapping or attaching heparin to the artificial cells by other means (Chang *et al.*, 1967b). BHC-collodion can also be used to directly coat particulate matter like activated charcoal (Chang *et al.*, 1968).

Effect Of Artificial Cells On Clotting

Blood was collected without stasis in 1 ml samples from a dog's femoral artery into 5 ml siliconized tubes, each containing 0.1 ml of artificial cells. Clotting time was recorded by a Lee-White procedure modified for siliconized tubes. The results given in Table III show that collodion or nylon membrane artificial cells accelerated the clotting of fresh blood in siliconized tubes, whereas BHC-collodion membrane artificial cells or nylon membrane artificial cells coated with BHC-collodion delayed clotting indefinitely. Blood kept in contact with one-tenth its volume of BHC-collodion artificial cells for more than one minute remained incoagulable indefinitely after removal of the artificial cells. Plasma from this blood did not clot when it was treated with normal oxalated plasma and then recalcified as in Wintrobe's screening test, though it did clot in the normal time when treated with the appropriate quantity of protamine sulfate. Similar results were obtained with nylon membrane artificial cells coated with BHC-collodion. The anticoagulant effect of the BHC-containing membranes in these *in vitro* tests could thus be ascribed, at least in part, to the release of an anticoagulant material, presumably heparin, from the artificial cells by contact with blood.

It could now be asked whether the anticoagulant effect of the BHC-collodion and BHC-collodion-nylon artificial cells was wholly due to the heparin leached out from them to attain a high concentration in the bulk phase, or whether there was in addition a nonthrombogenic effect at the surfaces. The following *in vivo* experiment was designed to test this point.

TABLE III
CLOTTING TIME OF BLOOD IN CONTACT WITH ARTIFICIAL CELLS

Type Of Artificial Cell	Clotting Time (Min.)	Number Of Dogs
Control blood	9.4 ± 3.5	10
Collodion	7.0 ± 2.5	9
BHC-collodion	>1500	9
Nylon	5.6 ± 2.0	5
BHC-collodion coated nylon	>1500	3

From Chang *et al.* (1967b). Courtesy of the National Research Council of Canada.

The shunt chamber (Fig. 48) was made from a standard blood administration pump set (R-78 Plexitron) by removing the ball valve, replacing the inlet tube with plastic tubing of the same type as the outlet tube, and coating the whole inner surface of the set, including the woven plastic filter, with graphite-BHC by the technique of Gott *et al.* (1964). To ensure that the surface was uniformly nonthrombogenic, the period of exposure to each of the coating liquids (graphite suspension, benzalkonium solution, and heparin solution) was doubled, and the heparin concentration was also doubled. The shunt chamber was rinsed thoroughly with saline before use to remove any free heparin. The application of graphite-BHC to the filter left it water-wettable and significantly narrowed its interstices.

Two of the shunt chambers described above were used in each experiment on dogs weighing about 20 kg and anesthetized with pentobarbital. The shunts were inserted between the femoral artery and vein on either side; one chamber contained 3 ml of BHC-collodion membrane artificial cells and the other 3 ml of ordinary collodion membrane artificial cells. The artificial cells of each lot had a mean diameter of about 100μ; those of less than 50μ diameter had previously been removed by passage through a filter similar to the one in the shunt. Arterial pressure was continuously monitored and remained in the range of 100 to 140 mm Hg. Except where clotting occurred, blood flow in each shunt was about 200 ml/min; at this flow rate the artificial cells within the filter did not impede the passage of blood through it. Blood from the efferent limb of each shunt and from a carotid artery was collected at 30-minute intervals into siliconized tubes, and its clotting time was measured.

The clotting time of blood samples obtained from the efferent limb of either shunt did not change significantly over two hours and was the same as that of blood from the carotid artery. With longer periods of insertion, blood flow over the control collodion membrane artificial cells gradually slowed, but flow over the BHC-collodion membrane artificial cells was well maintained. When the chambers were opened at the end of the test, small clots were found adhering to the artificial cells in the chamber containing the control collodion membrane artificial cells. In

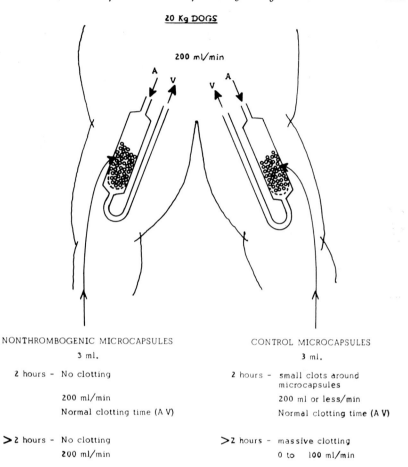

20 Kg DOGS

200 ml/min

NONTHROMBOGENIC MICROCAPSULES
3 ml.

2 hours - No clotting

200 ml/min

Normal clotting time (A V)

> 2 hours - No clotting

200 ml/min

CONTROL MICROCAPSULES
3 ml.

2 hours - small clots around microcapsules

200 ml or less/min

Normal clotting time (A V)

> 2 hours - massive clotting

0 to 100 ml/min

Figure 48. *In vivo* studies on the nonthrombogenicity of heparin-complexed artificial cells in nonthrombogenic extracorporeal shunt chambers perfused by recirculating blood.

some cases where blood flow had slowed, especially when perfusion was carried out for more than two hours, most of the collodion membrane artificial cells were trapped in large clots which extended downstream into the efferent limb of the shunt. In no case was any clotting observed in the chamber containing the BHC-collodion membrane artificial cells, provided there were no coating defects on its wall. These results show that BHC-collodion membrane artificial cells are nonthrombogenic, even when the blood flowing over them has a normal clotting time.

The BHC-collodion membrane artificial cells removed from the chamber after the experiment and tested *in vitro*, were found still to possess anticoagulant activity. These "used" artificial cells, when tested *in vitro*, yielded less anticoagulant in relation to the amount of blood in the bulk phase than did "fresh" artificial cells. Blood in contact with one-tenth its volume of fresh BHC-collodion artificial cells failed to clot if the artificial cells were removed after one minute of contact; on the other hand, a minimum of ten minutes' contact with used BHC-collodion artificial cells was required to make the blood incoagulable when the artificial cells were removed. Whether the anticoagulant activity of the BHC-collodion artificial cells could have been further reduced by longer exposure to flowing blood was not tested.

Blood in contact with BHC-collodion artificial cells might have failed to clot for one or more of three possible reasons: (1) the removal of a normal clotting factor (or factors) from the blood through destruction or adsorption by the relatively enormous surface area of the artificial cell membrane, (2) the leaching of heparin from the artificial cells in sufficient amount to make the blood in the bulk phase incoagulable, or (3) a specific non-thrombogenic property of the artificial cells. In the *in vitro* studies, the first possibility was excluded, and the second possibility was implicated by the inability of normal plasma and the ability of protamine sulfate to initiate clotting in plasma obtained from blood previously exposed to the artificial cells. It is clear that in these *in vitro* tests, where a blood sample was kept in contact with a relatively large volume of BHC-collodion artificial cells, the free heparin in the blood did reach the anticoagulant level. The third possibility, however, may have been responsible for the absence of clotting in the *in vivo* tests, in which the clotting time of the rapidly flowing blood was not altered by its brief passage over the artificial cells. The nonthrombogenic property of the BHC-collodion surface revealed by these *in vivo* tests could have been due either to the presence of heparin fixed on that surface, to the continuous removal of enough heparin to maintain an effective concentration in a thin layer of plasma immediately adjacent to the surface of the artificial cells, or to a combination of these two factors. It is of interest in this connection to note that heparin bound to BHC-graphite treated polymer rings placed in the inferior vena cava could exchange with circulating heparin and remain unclotted for 14 days (Whiffen and Gott, 1965). It has been suggested by the same authors that the heparin of the complex may be renewed by the endogenous heparin present in low concentration in normal plasma. A similiar mechanism may help to maintain a prolonged nonthrombogenicity of BHC-collodion artificial cells exposed to flowing blood.

Catalytic Activity Of Enzymes Enclosed In BHC-Collodion Artificial Cells

A model enzyme, urease, enclosed in BHC-collodion artificial cells (mean diameter 100μ) was found to catalyze the conversion of urea to ammonia as efficiently as urease enclosed in untreated collodion artificial cells of the same size (Chang *et al.*, 1967b). The encapsulated enzyme in either case was about 20 percent as active as the enzyme in free solution (Fig. 49). No leakage of enzyme from the artificial cells into the bulk phase could be detected. When 10 ml of such artificial cells was placed in the 70 ml shunt chamber described in an earlier paper (Chang, 1966), the chamber having been made nonthrombogenic by the procedure of Gott *et al.* (1964) and inserted into an arteriovenous shunt in an anesthetized dog, blood flow was well maintained over a 30-minute test period, although no anticoagulant solution was administered, and blood was propelled through the shunt only by the arterial pressure. During this time the blood ammonia level rose both in the shunt effluent and (more gradually) in the general circulation. Asparaginase enclosed in nonthrombogenic artificial cells also acted efficiently on its substrate (Chang *et al.*, 1968).

EFFECTS OF ARTIFICIAL CELLS ON FORMED ELEMENTS OF BLOOD

The relationship between the physicochemical properties of different types of artificial cell membranes and their effects on the formed elements of blood is of biophysical and physiological interest. At the same time, the selection of an artificial cell mem-

UREASE SOLUTION   UREA BUFFER
                                      ONLY

UREASE IN   UREASE IN BHC-COLLODION
COLLODION   MICROCAPSULES
MICROCAPSULES

O 10 20
MINUTES

Figure 49. Urease activity of urease solution compared to urease-loaded collodion membrane artificial cells and urease-loaded heparin-complexed collodion membrane artificial cells. (From Chang *et al.*, 1967. Courtesy of the National Research Council of Canada.)

brane which is least harmful to the formed elements of blood is important if the artificial cells are to be used in an extracorporeal shunt system. Experiments have been carried out to study the acute and long-term responses of animals to extracorporeal perfusion with three types of artificial cells (Chang *et al.*, 1968).

Artificial cells with nylon membranes, collodion membranes or heparin-complexed collodion membranes were prepared as previously described (Chang *et al.*, 1966, 1967b). Each type of artificial cell, 100μ mean diameter, contained hemolysate prepared from homologous erythrocytes. Each type of artificial cell was placed in the extracorporeal shunt chamber consisting of a chamber made from high density polypropylene (3 cm I.D. $\times$ 23.5 cm high), one steel-wire mesh and one nylon screen at each end, and two polyethylene cannulae each connected to a 16-gauge needle through a 3-way stopcock. All parts in contact with blood were coated with heparin-benzalkonium-graphite by Gott's procedure (1965).

In each experiment, shunts containing 10 ml of artificial cells were placed in the shunt and 4 litres of sterilized saline perfused through the shunt at 50 ml/min. During this time, direction of flow was reversed every 7.5 minutes. Finally the shunt primed with saline was connected to the animal's femoral artery and vein. The animal's blood pressure propelled blood through the shunt and the flow rate was adjusted to 20 ml/min. In order to facilitate mixing and to prevent packing of the artificial cells, it was essential to reverse the direction of blood flow every 7.5 minutes. When the reversal of flow was carried out at longer intervals, the artificial cells packed at the exit screen and adversely affected the platelet level of blood leaving the shunt. All flow was against gravity.

Aqueous heparin sodium (3 mg/kg) was injected intravenously when nylon or collodion artificial cells were used, but no heparin was given in the case of heparin-complexed collodion artificial cells. Arterial and effluent samples were taken simultaneously at 30-minute intervals for two hours. The shunt was disconnected after two hours, and during the postperfusion period arterial samples were taken at 30-minute intervals for another two hours.

Acute Experiment

During the two hours of perfusion through nylon membrane artificial cell shunts, there was a significant reduction in the arterial platelet level (Fig. 50). There was an even greater reduction in the platelet level in blood leaving the shunt. The arterial platelet levels remained low during the two-hour postshunt period, but had returned to the preperfusion value 24 hours later. Reduction in the leucocyte level was not observed. Instead, there was a significant increase in leucocyte count towards the last hour of the perfusion. This was followed by a very marked leucocytosis in the two-hour postshunt period (Fig. 51). There were no significant changes in hemoglobin or plasma hemoglobin.

During the two hours of perfusion through the shunts containing collodion membrane artificial cells, the platelet levels in

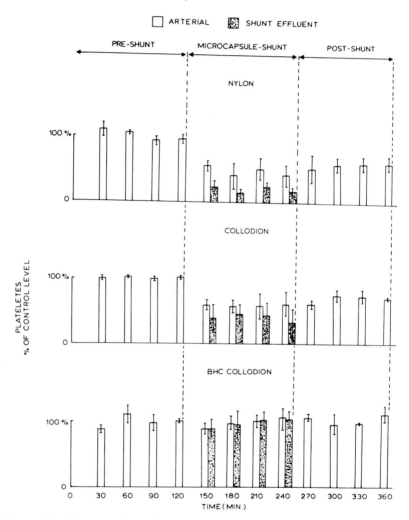

Figure 50. Effects of three different types of artificial cells on the platelet level of arterial and shunt-effluent blood. Each column represents mean and standard deviation of results obtained from three dogs. (From Chang *et al.*, 1968. Courtesy of the American Society for Artificial Internal Organs.)

the arterial samples and in the shunt-effluent samples were not reduced as much as in the case of nylon membrane artificial cells. However, unlike nylon membrane artificial cells, collodion membrane artificial cells caused a significant reduction in the

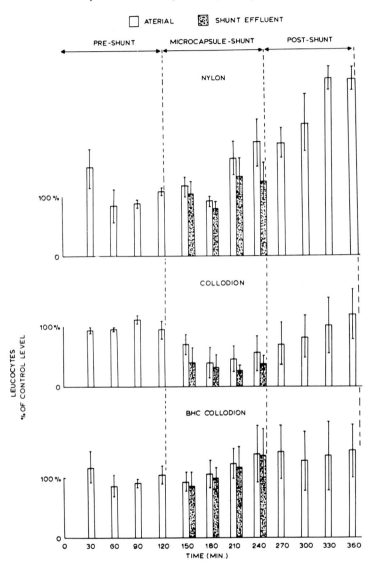

Figure 51. Effects of three different types of artificial cells on the leucocyte level of arterial and shunt-effluent blood. Each column represents mean and standard deviation of results obtained in three dogs. (From Chang *et al.*, 1968. Courtesy of the American Society for Artificial Internal Organs.)

leucocyte levels. The differential counts indicated that the major reduction was in the neutrophils and that this was especially so in the shunt-effluent samples. During the postperfusion period, the leucocyte level gradually returned towards normal, reaching the control level at the end of the two hour postperfusion period. There were no significant changes in hemoglobin or plasma hemoglobin.

The hematologic data remained very stable in the acute experiments using BHC-complexed collodion artificial cells. The stability of the platelet levels in the arterial and shunt-effluent samples was especially remarkable (Fig. 50). The leucocyte levels were stable, with a small but significant increase in level toward the end of the two-hour perfusion period (Fig. 51). There was a slight increase in plasma hemoglobin level (less than a 50% increase from the control preperfusion level). There were no changes in total hemoglobin, differential, or blood smears. Clotting times usually returned to normal two hours after the completion of perfusion and in all cases, 24 hours after perfusion.

Long-term Effects

The results of the acute experiments led to further investigations of the long-term effects of heparin-complexed collodion artificial cells. Results are summarized in Figure 52. The first points on the graph (day 0) represent control samples taken before perfusion. The lack of postperfusion anemia was shown by a slight but steady increase in the total hemoglobin level throughout the 147 days. The slight decreases in hemoglobin levels after each perfusion were due to the animals' blood lost in the extracorporeal shunt system, for only part of the animals' blood in the extracorporeal system was reinfused. The leucocyte levels remained fairly stable except for the 24-hour postperfusion samples. Leucocytosis was consistently observed 24 hours after the second, third, and fourth perfusions through the artificial cells. However, only one of the four dogs showed any leucocytosis after the first perfusion. It should be noted in this connection that whereas all the anmials received antibiotics (Fortimycin I.M.) after the first perfusion, antibiotics were purposely omitted after subsequent perfusions. The platelet levels fluctuated somewhat throughout the 147 days, but there was no consistent relationship between perfusions and fluctuations in platelet levels. The mean values of the four dogs were all above 200,000/mm^3. The mean clotting time of the four dogs taken 24 hours after each perfusion and thereafter, remained less than 20 minutes. (It should be noted that, as described in the acute experiments, sufficient heparin was leached out from the heparin-complexed collodion artificial cells so that clotting time of the animal did not return to normal until at least two hours after the completion of each perfusion.) Intradermal injections of heparin-complexed collodion artificial cells (washed free of traces of Tween 20) showed that repeated perfusions did not result in any hypersensitivity reactions.

Discussion

The effects of different types of artificial cells on the formed elements of blood is of considerable basic biophysical and physiological interest. Collodion membrane artificial cells cause a reduction of both platelet and leucocyte levels, but the complexing of a small amount of heparin-benzalkonium to the same membrane results in a surface which no longer has any significant effect on the formed elements of blood. Besides the presence of heparin on a membrane, other physicochemical factors are also important. Studies on the different types of heparin-complexed membranes show that formed elements of blood have less tendency to adhere to some types of heparinized surfaces (Leininger *et al.*, 1966) than other (Grode *et al.*, 1968). While the present study is rather limited, the results obtained show that heparin-complexed collodion artificial cells do not cause a significant reduction of the formed elements of blood. This is so in both acute and long-term experiments. While heparin-complexed collodion artificial cells have the least adverse effect on the formed elements of blood, nylon and collodion membrane artificial cells have other advantages. Nylon is a much stronger polymer, and its effect of decreasing platelets and causing leucocytosis is not much more pronounced than that of the artificial kidney (Lawson *et al.*, 1966), blood pumps, leucocytes (Kusserow *et al.*, 1965). Although extracorporeal shunts offer the most efficient

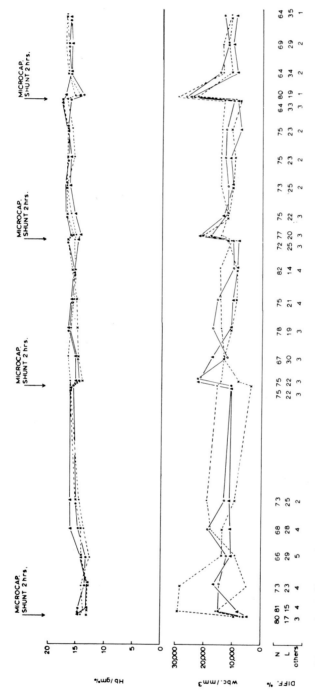

Figure 52. Effects of intermittent extracorporeal perfusion through heparin-complexed collodion membrane artificial cells. Each point represents the data from one dog. Only the results of the four dogs used in intermittent perfusions are shown. The other two dogs which were perfused once have similar postperfusion data. (From Chang *et al*, 1968. Courtesy of the American Society for Artificial Internal Organs.)

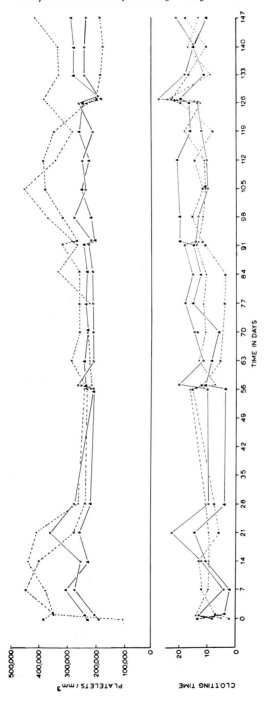

way for artificial cells to act on blood-borne substrates or toxins, other routs of administration are possible. Recirculation of peritoneal fluid through an artificial cell shunt, oral ingestion of artificial cells, parenteral administration, or recirculation of dialysis fluid of an artificial kidney through an artificial cell shunt have all been attempted in this laboratory.

Chapter 8

ARTIFICIAL CELLS CONTAINING DETOXICANTS

INTRODUCTION

SINCE the report by Muirhead and Reid (1948), much work has been done using hemoperfusion over resins to remove exogenous toxins or endogenous metabolites (cf. Schechter *et al.*, 1958; Pallotta and Koppanyi, 1960; Nealon and Ching, 1962; Nealon *et al.*, 1966; Rosenbaum *et al.*, 1970). Unfortunately, there are a number of problems preventing their widespread clinical use. When in direct contract with blood, ion exchange resins reduce blood platelet and leucocyte levels (Rosenbaum *et al.*, 1970), and remove essential elements like calcium ions from perfusing blood (Rosenbaum *et al.*, 1962). Another class of detoxicants which has been widely studied is activated charcoal. It has been known for many years as a useful detoxifying agent. For many years it has been used orally to remove ingested toxins in the gastrointestinal tract. However, activated charcoal administered orally does not act with sufficient speed or efficiency in patients who have already absorbed a large amount of toxin and are deeply intoxicated. In these cases, it has been found that a column of granular activated charcoal can efficiently remove the toxins from perfusing blood. Thus barbiturates (Yatzidas, 1964; Hagstram *et al.*, 1966), Doriden® (DeMyttenaere *et al.*, 1967), and other toxins (Dunea and Kloff, 1965) have been successfully removed from a blood-perfusing activated charcoal column. Unfortunately, we are again faced with a number of problems which prevent the widespread clinical use of hemoperfusion over activated charcoal. Activated charcoal in direct contact with blood adversely affects platelets (Dunea and Kolff, 1965; Hagstam *et al.*, 1966; DeMyttenaere *et al.*, 1967; Dutton *et al.*, 1969). In addition, the charcoal granules, despite careful washing, release free particles into the circulation, causing emboli in the lungs, liver, spleen, and kidney (Hagstam *et al.*, 1966).

The principle of artificial cells has been used to circumvent these problems (Chang, 1966) (Fig. 2). Detoxicants enclosed in artificial cells will not come into direct contact with the formed elements of blood. Powder released from the enclosed detoxicant cannot leave the artificial cells to cause embolism. In addition, the enclosing membranes can be made selectively permeable to toxins so that essential elements of blood like calcium can be excluded. On the other hand, external permeant toxins can diffuse freely across the artificial cell membranes to be removed by the enclosed detoxicants. The selective removal of toxin can be accomplished by variations in membrane porosity, lipids, charge, and other properties.

This study has been carried out through a number of stages. Initial work was done (Chang, 1966; Chang *et al.*, 1967b) using a slight modification of the standard procedure (Chang, 1964; Chang *et al.*, 1966) to enclose ionexchange resins and activated charcoal in artificial cells. When this basic research demonstrated the feasibility of this system, modified procedures were used to prepare artificial cells suitable for use in scaled-up systems (Chang *et al.*, 1968, 1970; Chang, 1969e), which is now being tested clinically (Chang, *et al.*, 1970, 1971).

STANDARD ARTIFICIAL CELLS

With slight modification of the standard procedure, nylon artificial cells containing Dowex®-50W-X12 have been produced (Fig. 53). Ion exchange resins are suspended in the aqueous phase containing hemoglobin and alkaline diamines. After adjusting the pH to 10.5, the remaining steps for the preparation of nylon membrane artificial cells were followed. Shunts have been loaded with these artificial cells of up to 120 ml. Over short periods of perfusion, such a shunt is effective in removing most of the ammonia from the perfusing blood of animals whose ammonia levels have been raised by the continuous infusion of ammonium sulfate. Some technical difficulties still have to be overcome before this system can be used for prolonged perfusion. In particular, the artificial cells, whose membranes are flexible, tend to become so closely packed when used in such large volume that blood flow diminishes in spite of repeated reversal of flow. Activated charcoal can also be microencapsulated

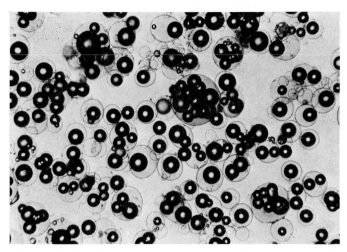

Figure 53. Artificial cells containing enzymes and Dowex-50W-12. (From Chang, 1966. Courtesy of the American Society for Artificial Internal Organs.)

(Chang, 1966). This is done by suspending activated charcoal powder in the aqueous phase, followed by the standard procedure for the preparation of artificial cells with nylon membranes, collodion membranes, or heparin-complexed membranes (Chang, 1964; Chang *et al.*, 1966, 1967b). Other laboratories have also been successful in using our procedures for the enclosure of ion exchange resins or activated charcoal in nylon membrane artificial cells (Levine and LeCourse, 1967; Sparks *et al.*, 1969).

Because of the flexibility of the membranes, this system, though useful for small-scale basic research, cannot be readily used in scaled-up systems suitable for clinical use. For this, further steps of development are required.

BHCAC ARTIFICIAL CELLS

Having demonstrated that polymer coating can prevent the problems of packing and breaking of membranes under a high flow situation, heparin-complexed collodion was used as the enclosing membrane (Chang *et al.*, 1968), since earlier results (Chang *et al.*, 1967b, 1968) indicated that it is the least harmful to the formed elements of blood. Coating of particles can be done in a large variety of ways. For instance, in a typical example (Chang *et al.*, 1968), 100 ml of heparin-benzalkonium collodion solution prepared as described (Chang *et al.*, 1967b) is added to 50 gm of activated charcoal. (Instead of the smaller granules, the 6-14 mesh granules, Fisher Co., are less liable to cause packing.) The suspension is manually stirred until the solvent of the polymer solution has evaporated sufficiently so that the microencapsulated particles are no longer adherent to one another. The artificial cells are then spray-dried at room temperature and suspended in saline over night. Experiments have been carried out to test the effect of microencapsulated activated charcoal on the formed elements of blood. Forty grams of either the uncoated activated chracoal or the microencapsulated activated charcoal formed by coating with heparin-complexed collodion was placed in the extracorporeal shunt chamber described. Blood from dogs was perfused through the shunt at 100 ml/min. The effect on platelet level is shown in Figure 54. Activated charcoal enclosed in artificial cells did not have any adverse effect on the blood platelet level. On the other hand, free activated charcoal which was not enclosed in artificial cells caused a 50% reduction in the arterial platelet level of heparinized dogs and an even greater decrease in the shunt effluent level. The use of the large mesh activated charcoal granules (6-14 mesh) allowed a large amount of heparin-complex coated activated charcoal to be used (300 gm). In addition, a blood flow rate of 200 ml/min or more could be used without resulting in packing at the exit screen.

This system does not adversely affect the formed elements of blood and does not result in the release of embolising powder. It is also capable of removing some Doriden, salicylate, Nembutal®, and uremic metabolites. However, the membrane thickness prevents the efficient removal of these toxins. With thinner membranes this preparation is more efficient, but in this form, the membrane contained too little heparin to have a platelet protection and nonthrombogenic effect. This problem is being solved by variation in membrane materials and changes in heparinization procedures. In the meantime, a different approach has led to the preparation of a system for clinical use.

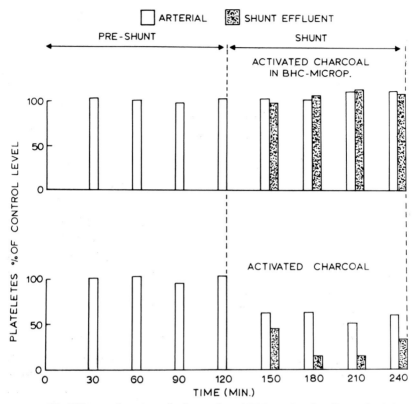

Figure 54. Effects of activated charcoal in BHC-artificial cells and of free activated charcoal on the platelet level of arterial and shunt-effluent blood. (From Chang *et al.*, 1968. Courtesy of the American Society for Artificial Internal Organs.)

ACAC ARTIFICIAL CELLS

In this approach, instead of using heparin, albumin is complexed into the artificial membrane by the following procedure (Chang, 1969e): 400 gm of activated charcoal granules (6-14 mesh, Fisher Co.) wrapped in sterile cloth is autoclaved at 121°C, 15 pounds per square inch, (psi) for one hour. After this, it is left in the 50°C ventilator for at least 96 hours, A polymer solution is made up by dissolving 20 ml of collodion (U.S.P.) in an organic solvent consisting of 20 ml of alcohol and 400 ml of ether. The 440 ml of polymer solution is poured into a beaker containing the 400 gm of activated charcoal. The suspension is stirred manually with a metal stirring rod until a free-flowing polymer solution is no longer present in the container. The slightly wet polymer-coated activated charcoal is spread out in a 2x1-foot tray. This is then placed in a ventilated oven for five hours at 50°C. After this the polymer-coated activated charcoal granules are placed in 10 liters of pyrogen-free distilled water. The suspension is sieved (No. 20 mesh) continuously with more pyrogen-free water until all fine particles are removed. The 400 gm of polymer-coated acti-vated charcoal is then placed in a silicone-coated high density polypropylene shunt prepared as described (Chang, 1969e). The shunt is then drained, stoppered, wrapped, sealed and placed in the autoclave for 30 minutes at 121°C, 15 psi. It is important not to pass through the stage of "liquid cool"; instead, the auto-clave is "vented" at the end of 30 minutes. At the end of this, the shunt is allowed to cool in its sealed, sterilized wrapping. Using sterile technique, the shunt is perfused with 4 liters of saline (Baxter, for human use) at 200 ml/min by gravity. A 1 gm/100 ml albumin solution in sterilized saline is prepared from human albumin (Cutter Co.). This albumin solution is used to displace all the saline from the shunt containing the polymer-coated acti-vated charcoal. The shunt is then sealed by sterilized technique and kept at 4°C for 15 hours for the albumin to coat the polymer membrane. Just before use, the albumin solution is displaced with 4 liters of saline (Baxter, for human infusion) without allowing the albumin-polymer-coated activated charcoal to come into contact with air at any time. After injecting 2000 units of heparin into the shunt, the system is now ready for use by attachment to the arteriovenous shunt.

Hematological Findings: Formed Elements Of Blood

Detailed studies of different types of microencapsulated acti-vated charcoal on the formed elements of blood have been done (Chang, 1969e; Chang and Malave, 1970). In these experiments 3 mg/kg of heparin was used, 1 mg/kg into each shunt containing 300 gm of ACAC microencapsulated charcoal and 2 mg/kg intravenously. After taking control samples, hemoperfusion (120 ml/min) was started; no blood pump was used. The results ob-

TABLE IV
EFFECTS OF 2-HOUR HEMOPERFUSION ON SYSTEMIC
PLATELETS AND LEUCOCYTES

| | Platelets/mm³ | | WBC/mm³ | |
| | | 2 hr | | 2 hr |
Dogs	Initial	Hemoperfusion	Initial	Hemoperfusion
		CONTROL		
1	141,000	129,000	16,700	21,000
2	198,000	209,000	10,800	22,000
3	128,000	131,000	8,360	15,000
4	183,000	201,000	10,000	10,400
		ACAC MICROENCAPSULATED ACTIVATED CHARCOAL		
5	310,000	301,000	23,000	26,000
6	260,000	250,000	13,000	11,800
7	383,000	334,000	11,500	13,970
8	345,000	295,000	12,900	9,900
9	108,000	125,000	2,900	2,800
10	317,000	245,000	5,100	6,000
11	260,250	201,000	8,000	7,100
12	211,000	179,000	6,800	9,900
13	157,000	123,000	4,850	3,900
14	294,000	218,000	19,000	13,000
15	305,000	249,000	11,000	14,000
16	106,000	106,000	3,300	2,500
		FREE ACTIVATED CHARCOAL		
17	208,000	68,000	11,700	3,200
18	130,000	78,000	6,900	2,800
19	169,000	92,000	6,600	1,500
20	273,000	119,000	8,700	3,980

From Chang and Malave (1970). Courtesy of the American Society of Artificial Internal Organs.

tained are briefly summarized in Table IV. In hemoperfusion across free activated charcoal granules there was a marked drop in systemic platelets and leucocyte levels. ACAC microencapsulation prevented this adverse effect on platelets or leucocytes. Plasma hemoglobin measurements showed that neither free activated charcoal nor ACAC microencapsulated activated charcoal caused any hemolysis of perfusing blood.

Histological Findings

After hemoperfusion through free activated charcoal, Hagstam (Hagstam *et al.*, 1966) found charcoal particles in effluent blood and masive charcoal emboli in the lungs, livers, kidneys, and spleens of the test animals. In our experiments using artificial cells containing activated charcoal (ACAC artificial cells), no charcoal particles were found in smears of effluent blood from shunts (Chang, 1969e); however, charcoal power was consistently observed from those shunts containing free activated charcoal (Chang, 1969e). Studies have been done on histological sections of lungs after two hours of hemoperfusion through 300 gm of ACAC microencapsulated activated charcoal (Chang and Malave, 1970). Sections obtained from four dogs were examined carefully and no charcoal emboli could be found in the first three dogs treated with preparations which had not been autoclaved. In the fourth dog treated with autoclaved microencapsulated charcoal, four very small charcoal particles in the precapillary vessels were found. With more careful washing of the autoclaved preparations, no charcoal particles were found in the histological sections of the fifth, sixth, and seventh dogs. Histological sections of liver, spleen, and kidney obtained from four dogs showed no evidence of charcoal embolism. Thus, unlike the massive embolism resulting from hemoperfusion through free activated charcoal, hemoperfusion through carefully washed ACAC artificial cells did not result in any such embolism.

Long-Term Follow-Up Studies

In control experiments, four Nembutal-anesthetized dogs underwent percutaneous femoral artery catheterization (14-gauge catheter). In hemoperfusion experiments, seven Nembutal-anesthetized dogs underwent similar percutaneous arterial catheterization. Each of these animals was then treated with two hours of hemoperfusion through 300 gm of sterile ACAC microencapsulated activated charcoal and then allowed to recover. All animals have been followed for more than one month. There were no adverse effects.

Pentobarbital As A Model Exogenous Toxin

The efficiency of the ACAC artificial cell system for the removal of exogenous toxins has been tested using a barbiturate, pentobarbital, as a model system (Chang 1969e, 1971b; Chang *et al.*, 1970). Barbiturates belong to the group of drugs commonly encountered in accidental or suicidal overdosage. In these studies dogs are given intravenous injections of pentobarbital (50-60 mg/kg) and then supported by artificial respiration. Control studies show that the rate of decrease of the arterial pentobarbital level is slow. On the other hand, hemoperfusion over ACAC artificial cells rapidly lowers the arterial pentobarbital levels. A typical example is shown in Figure 55. Forty-five minutes after the infusion of pentobarbital, control blood samples are taken at 15-minute intervals for one hour. An extracorporeal shunt containing 300 gm of ACAC artificial cells is then connected to the femoral artery and femoral vein of the heparinized animal. A Travenol blood pump is used to maintain an extracorporeal blood flow of 200 ml/min. As can be seen in Figur 55, the arterial blood pentobarbital level fell exponentially and rapidly during the two hours of hemoperfusion. The effectiveness of the hemoperfusion over ACAC artificial cells is shown both by the rate of decrease of arterial pentobarbital levels and also, in some cases, by the animal's waking up from the effects of the pentobarbital.

Pentobarbital in the body is distributed into four compartments: blood pool, viscera, lean tissues, and adipose tissues (Dedrick and Bischoff, 1966). The visceral compartment, which includes the central nervous system, receives a high blood flow, whereas the adipose tissue receives a low blood flow. Thus, using this compartmental system, we can say that hemoperfusion over ACAC artificial cells removes pentobarbital from the blood compartment along with any pentobarbital entering the blood compartment along a concentration gradient from the other compartments. From the analysis of Dedrick and Bischoff (1966), it can be seen that the rate of fall of free concentrations of pentobarbital in the visceral compartment (including the central nervous system) falls at the same rate as the pentobarbital concentration in the blood compartment; on the other hand, the concentration

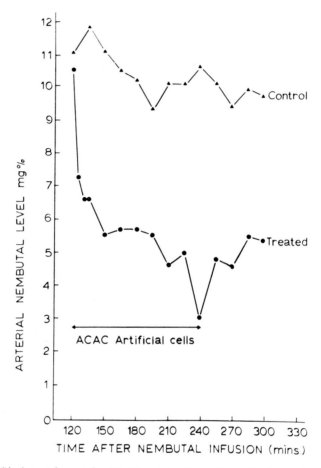

Figure 55. Arterial pentobarbital level in dogs. Control: dog without treatment. Treated: dog with systemic circulation connected to an extracorporeal shunt containing 300 gm of ACAC artificial cells (blood flow 200 ml/min) for two hours.

in the adipose tissue lags behind the other compartments. Thus it would be expected that hemoperfusion over ACAC artificial cell could efficiently remove pentobarbital from the blood compartment and the central nervous system sufficiently to revive the comatose individual. On the other hand, on termination of hemoperfusion, pentobarbital will continue to diffuse along a concentration gradient from the adipose tissue into the blood compartment, thus resulting in a gradual increase of the blood pentobarbital level, as can be seen in Figure 55. The degree of increase of blood pentobarbital level will depend on the amount of adipose tissue of any one particular individual.

Salicylates As A Model Exogenous Toxin

Aspirin, a salicylate, is the most important cause of accidental poisoning among children under the age of five years. The efficiency of ACAC artificial cells in the removal of salicylate has been studied in this laboratory using dogs.

A typical example is shown in Figure 56. In each set of experiments 200 mg/kg of sodium salicylate was given intravenously by infusion to two Nembutal-anesthetized dogs which had previously been nephrectomized to prevent the renal excretion of salicylates. Arterial salicylate levels were followed at 15-minute intervals for one hour. After this, one of the dogs was used as a control. The circulation of the other dog was connected to an extracorporeal shunt containing 300 gm of ACAC artificial cells. Hemoperfusion over the ACAC artificial cells in the extracorporeal shunt was facilitated by the use of a Travenol rotary pump adjusted to a flow rate of 200 ml/min. The arterial salicylate levels in the control animal remained elevated. During this period, the animal exhibited the usual signs of salicylate intoxication. The period of marked hyperpnea was followed by respiratory failure, cardiovascular collapse, and death of the untreated animal. In the case of the animal treated with the ACAC artificial cells, the arterial salicylate levels fell rapidly, and the clinical signs of salicylate intoxication were no longer present soon after the initiation of hemoperfusion. Clearance of salicylate was maintained at 110 ml/min to 120 ml/min for the first 90 minutes.

The efficiency of the ACAC artificial cells for the treatment of salicylate intoxication is clearly demonstrated in these experiments. The size of the animals (25-30 kg) and the amount of salicylates administered (4.5-5.5 gm to each animal) in these studies are comparable to what one may encounter in pediatric salicylate intoxication.

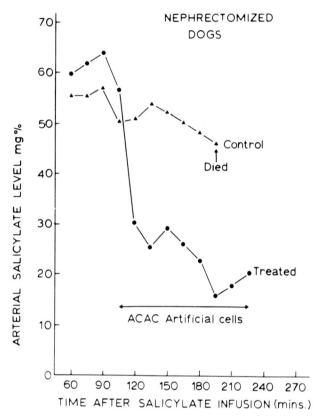

Figure 56. Arterial salicylate levels in dogs. Control: dog without treatment. Treated: dog with systemic circulation connected to an extracorporeal shunt containing 300 gm of ACAC artificial cells (blood flow 200 ml/min) for two hours.

Doriden As A Model Exogenous Toxin

Doriden (glutethimide) is another medication commonly encountered in acute intoxication. Preliminary studies in this laboratory using dogs show that Doriden is efficiently removed by the ACAC artificial cells. Dogs were given 150 mg/kg of Doriden. The animal's blood was recirculated at 300 ml/min (rotary pump) through the extracorporeal shunt chamber containing 300 gm of ACAC artificial cells. This way the clearance of Doriden was found to be 150 ml/min for the first hour and

120 ml/min for the second hour. Thus, about 1 gm of Doriden was removed by the 300 gm of ACAC artificial cells within two hours.

GENERAL CONCLUSIONS

This chapter serves to illustrate the feasibility of the general idea of artificial cells. By placing detoxicants in an artificial intracellular environment, the detoxicants can continue to remove permeant toxins without adversely affecting extracellular essential elements like blood cells, protein, and calcium. The examples given in this chapter serve only as model systems to test the efficiency of the artificial cell system. The demonstrated efficiency of these model systems should encourage one to study artificial cells containing other types of detoxicants for the removal of the large numbers of exogenous and endogenous toxins. The simplicity and the compactness of this system will greatly facilitate the treatment of acute intoxication. Other membrane systems should also be explored. Thus the Utah group (Kolff, 1970, and Andrade *et al.*, 1971a and 1971b) is now exploring the microencapsulation of activated charcoal by coating with Hydron, a polymer known for its biological inertness. We are also exploring other membrane systems including cellulose acetate in this laboratory.

Chapter 9

ARTIFICIAL CELLS FOR ARTIFICIAL ORGANS

INTRODUCTION

SINCE cells are the fundamental units of all tissues and organs, what are the possibilities of using the principle of artificial cells for the construction of artificial organs? The idea of artificial cells has been applied to the design and construction of compact artificial kidneys (Chang, 1966), with the hope that this can be used as a feasibility test for the construction of other types of artificial organs. Part of this system is now being tested clinically with promising results (Chang and Malave, 1970; Chang *et al.*, 1971).

CONVENTIONAL ARTIFICIAL KIDNEYS

Before discussing the details of artificial kidneys constructed from artificial cells, a brief summary of the basic functional characteristics of the conventional artificial kidneys should be given. It is obviously not within the scope of this monograph to review meaningfully the large amount of work on artificial kidneys, nor is it possible to include references for all the many publications on this subject. A few of the many excellent publications include monographs (Kolff, 1946; Schreiner and Maher, 1961; Doyle, 1962; Merrill, 1965; Hampers and Schupak, 1967; Brest and Moyer, 1967; Nose, 1969), journals (Schreiner, 1955-1970; Kerr, 1967-1970), reviews (Leonard, 1966; Dedrick *et al.*, 1968), and reports (Burton, 1967; Gottschalk, 1967).

In 1913, Abel *et al.* devised a hemodialysis apparatus to remove toxic materials in experimental animals. Kolff and Berk in 1943 developed the rotating drum artificial kidney and demonstrated for the first time the clinical usefulness of hemodialysis. With the development of the Scribner-Quinton arteriovenous shunts (Quinton *et al.*, 1960), long-term hemodialysis became a reality. Other centers, realizing the importance of these findings, have placed increasing emphasis on the development and uses of hemodialysis.

146

At present, different types of artificial kidneys are being used. These include the coil type, e.g. Kolff twin coil artificial kidney (Kolff and Watschinger, 1956); Alwall coil artificial kidney (Alwall, 1947); the plate type, e.g. Skeggs and Leonard's (1948) artificial kidney; the Kiil (1960) type artificial kidney; and more recently, the capillary type (Stewart *et al.*, 1966). All these artificial kidneys have in common the following basic principle (Fig. 57): a semipermeable membrane separates two compartments. The patient's blood flows through one compartment. Here, permeant molecules from the patient's blood cross the semipermeable membrane by diffusion along electrochemical and hydrostatic gradients to the other compartment where they are washed away by a very large volume of dilaysate (200-300 liters). The rate of removal of waste metabolites depends on a number of factors. The membrane thickness and total membrane area available for diffusion are two of the factors. The success of this basic principle is clearly demonstrated by the large number of chronic renal failure patients successfully maintained by chronic hemodialysis. Despite the therapeutic value of hemodialysis, Professor Kolff has recently quoted that only 1.5 percent of the people who need it are being helped with artificial kidney or kidney transplantation (Nose, 1969). In other words, in the United States alone, out of 40,000 people who require treatment only about 500 can be accommodated. Serious attempts are being made by many centers to construct simpler and less costly artificial kidneys.

THEORETICAL ASPECTS OF ARTIFICIAL KIDNEYS BASED ON ARTIFICIAL CELLS

The basic principle of the use of artificial cells for the construction of compact artificial kidneys was first proposed in 1966 (Chang, 1966) (see Fig. 57). It is well known that the smaller a particle, the larger is the surface area to volume relationship. If we calculate the total surface area in artificial cells of different diameters, the result obtained is quite striking. Thus, 10 ml of 20μ diameter artificial cells or 33 ml of 100μ diameter artificial cells has a total surface area about 20,000 cm^2, which is larger than that available in a conventional artificial kidney (less than

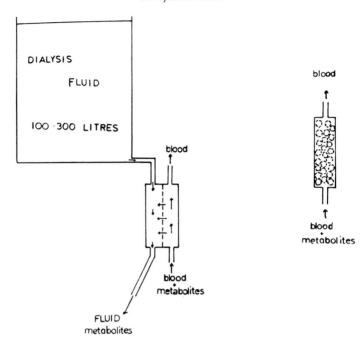

TOTAL SURFACE AREA OF MEMBRANE

Artificial kidney	Microcapsules
20,000 cm²	20,000 cm²
	10ml (20μ)
	33ml (100μ)

Figure 57. Schematic representation of principle of conventional artificial kidney (hemodialyzer) (*left*), and artificial cells for artificial kidney (*right*). (From Chang and Malave, 1970. Courtesy of the American Society for Artificial Internal Organs.)

20,000 cm²). In addition, the membrane thickness of artificial cells (less than 500 Å) is much less than that used in standard artificial kidneys (at least 50,000 A). If everything else remains the same, the rate of net movement of permeant molecules across a membrane (ds/st) along a given concentration gradient (ΔCs) is proportional to the surface area (A) and inversely proportional to the membrane thickness (dx), since

$$ds/dt = PA\frac{\Delta Cs}{dx}$$

The principle of the construction of a compact artificial kidney from semipermeable artificial cells is simply as follows: If we pack 33 ml of 100μ diameter artificial cells in a shunt, then the total membrane area available for diffusion will be greater than that of an artificial kidney. Since the membrane thickness is at least two orders of magnitude less than that in the standard artificial kidney, metabolites from blood flowing past these artificial cells can cross the membrane into the artificial cells at least 100 times faster than in standard artificial kidneys. If something can be placed inside these artificial cells to trap or act on the metabolites which cross the membrane, then we have the theoretical basis for a small, compact artificial kidney. For this model to become practical, we have to answer the following questions: What are the permeability characteristics of the artificial cell membrane? What is the compatibility of artificial cells with blood? What can be included within the artificial cells to trap waste metabolites?

The permeability characteristics of artificial cells have been studied in detail (Chang, 1965; Chang and Poznansky, 1968c). As discussed in the earlier chapter, the equivalent pore radius of the artificial cells can be varied at will over a wide range. Those prepared by the standard procedures (Chang, 1964; Chang *et al.*, 1966) have an equivalent pore radius of 18 Å, a value comparable to that used in dialysis membranes. Thus, the artificial cell membranes are impermeable to particulate matters (e.g. blood cells or suspensions) and macromolecules (e.g. enzymes and other proteins). Smaller molecules, however, can equilibrate rapidly across the membranes. The rate of equilibration of uremic metabolites like urea, creatinine, and uric acid is so rapid that a mixing and sampling procedure is required for the analysis of permeability constants. The results obtained and described in an earlier chapter show that the permeability characteristics of artificial cell membranes are suitable for the rapid permeation of dialyzable uremic metabolites. In addition, the permeability characteristics of artificial cell membranes can be modified by variations in polymer materials, thickness, porosity, charge, lipid coating, and polysaccharide coating.

Results described in the previous chapter indicate that artificial cells can be prepared to have membranes which are compatible with the formed elements of blood. The next question which requires solution is the type of materials for the removal of uremic metabolites.

REMOVAL OF BLOOD UREA

At present, no absorbent is available which has a sufficient absorbing capacity for the removal of blood urea. Studies in this laboratory demonstrated the possibility of using artificial cells containing urease (Chang, 1966).

Procedures And Methods

Control artificial cells containing no urease and urease-loaded artificial cells (soluble urease, Nutritional Biochemical Co., 400 mg/3.0 ml artificial cells) are prepared according to the standard procedure described for nylon artificial cells of 90μ mean diameter (Chang *et al.*, 1966). Before use, Tween 20 must be removed as completely as possible from the artificial cell suspension by washing at least six times on the centrifuge with 10 volumes of saline.

The extracorporeal shunt system I have used in experiments on dogs is shown diagrammatically in Figure 46. The artificial cells (90μ mean diameter, smallest ones removed by differential screening) must be retained in the shunt chamber without stasis or close packing and without interfering with the free perfusion of blood. This is accomplished by placing a 325-mesh nylon screen, supported by a steel wire screen, at either end of the chamber; the distance between the screens is 15 cm and the internal diameter of the chamber is 2.0 cm. With the pump adjusted to give a perfusion rate of 20 ml/min, it is desirable to reverse the direction of flow every 10 minutes to prevent clogging of the screen. The system functions well for three to four hours; for longer periods of operation the shunt chamber needs further modifications, or the perfusion may be switched to a second chamber.

Nembutal-anesthetized dogs with an average weight of 10 kg were used in the acute experiments. Right femoral artery (A) and vein (V) were cannulated and connected to the shunt system (Fig. 46). Each dog received 2.0 mg/kg of aqueous heparin sodium (Nutritional Biochemical Co.) intravenously. In each case, arterial blood was perfused for 90 minutes through the shunt chamber; during this period the chamber contained 10 ml of control nylon artificial cells which were not loaded with urease. Thereafter, arterial blood was perfused for a second 90-minute period through the same chamber, which then contained 10 ml of urease-loaded artificial cells.

Results

Systemic blood ammonia levels were never significantly changed when arterial blood was perfused at 20 ml/min through a shunt containing control artificial cells. On the other hand, systemic blood ammonia levels were regularly increased when the shunt chamber contained urease-loaded artificial cells. Thus blood ammonia rose steeply, the average level after 90 minutes being 28.5 (range 20.0-34.0)μg/ml (Fig. 58), in spite of the presence in these dogs of a functioning liver which was presumably converting ammonia to urea during the course of the experiment. Ammonia in the effluent was too high and urea in the effluent was too low for satisfactory estimation by the methods routinely employed, but it was clear that at least 80 percent of the urea in the blood perfusing the shunt had been converted to ammonia. The results for ammonia and urea in systemic arterial samples from this experiment are shown in Figure 58, where the values are also plotted in terms of urea and ammonia N. The rise in blood ammonia N accounted for about three-fourths of the fall in blood urea N; a closer agreement would hardly be expected in view of the likely differences in distribution and in renal and metabolic handling of the two substances.

The urea-urease system has been used here as a model to test the feasibility of using urease-loaded artificial cells to lower blood urea level. From this point of view the results seem promising; the shunt gave more than 80% conversion, though the volume of the incorporated artificial cells was only 10 ml. That ammonia formed from the enzymatic action on urea could be removed by artificial cells containing ammonia absorbent has

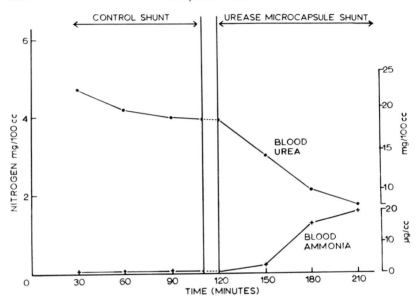

Figure 58. Extracorporeal shunt containing urease-loaded artificial cells. Effect on systemic arterial blood ammonia and blood urea level. (From Chang, 1966. Courtesy of the American Society for Artificial Internal Organs.)

been demonstrated in the same studies (Chang, 1966). Thus with slight modification of the procedure, nylon artificial cells containing Dowex-X12 can be produced. Shunts have been loaded with artificial cells of up to 120 ml. Over short periods of perfusion such a shunt chamber was effective in removing most of the ammonia from the perfusing blood of animals whose ammonia levels have been raised. For practical use, ammonia absorbents with much higher capacity will be required.

General Discussion

This study demonstrated the theoretical possibilities of using artificial cells containing a combination of urease and ammonia absorbents for the removal of blood urea. Other workers have supported the feasibility of using artificial cells for the removal of urea. Thus, Levine and LaCourse (1967) analysed the theoretical feasibility of a column of artificial cells containing urease to remove blood urea. In this analysis, the extra renal blood volume and urea concentration are represented respectively by V_B and U_B, and the volume of blood flowing through the artificial kidney per unit time is represented by V. If M is the rate of urea production and Uo the concentration of urea at the efferent of the shunt, then

$$V_B(dU_B/dt) = V(Uo-U_B) + M$$

If f represents the ratio of Uo/U_B, then the steady-state solution to the above equation becomes

$$U_B = M/V(1-f)$$

Thus the steady-state concentration depends on V, the rate of blood through the artificial kidney, on M, the rate of urea production, and on f, the ratio between the urea concentration in the afferent (U_B) and efferent (Uo) of the artificial kidney. If f is very much less than 1 (in fact about 0.1, as shown experimentally by Chang, 1966), and knowing M (25 gm/24 hr and U_B (0.25 gm/liter), then the required blood flow rate through the artificial kidney can be estimated at about 70 ml/min.

With further derivations taking into account enzyme kinetics, membrane permeability, blood flow rate, and other factors, the authors (Levine and LaCourse, 1967) arrived at the following equation,

$$h = \frac{v \ln 1/f}{3PF} r$$

where h is the length of column required, v is the linear velocity of blood, f is the ratio of the efferent urea concentration, P is the permeability constant of the artificial cell membrane, F the fraction of the column occupied by artificial cells, and r the radius of the artificial cells. They used this equation to derive a graph (Fig. 59) showing column height as a function of arterial cell diameters for two values of the permeability constant. This graph shows that in theory, it should be possible to use artificial cells to construct an artificial kidney 2 cm in diameter and 10 cm in length for the conversion of urea produced daily to ammonia.

Our approach of combining urease and ammonia absorbents for the removal of blood urea has been further extended. It has been suggested (Chang, 1966) that by placing such materials as these in the external dialyzing fluid of an artificial kidney, the

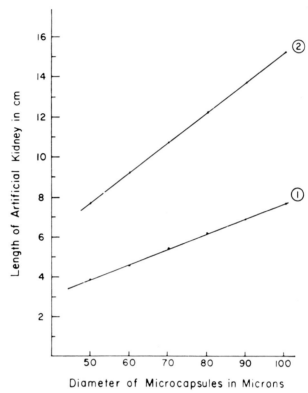

Figure 59. The length of column required for f = 1/10 is plotted versus artificial cell diameter. P = 1 × 10⁻⁴ cm/sec in curve 1, and P = 5 × 10⁻⁵ cm/sec in curve 2. Other details are described in the test. (From Levine and LaCourse, 1967. Courtesy of Interscience Publishers, New York.)

urease Dowex system can also act on urea diffusing across the hemodialyzer membranes. The main advantage of enclosing them in artificial cells is the enormous ratio of surface area to volume that may be attained in this way. With artificial cells of 20μ diameter, for instance, the surface area available for exchange is 2,500 cm²/ml of artificial cells. Thus 10.0 ml of such artificial cells have more surface area than a conventional artificial kidney. This large surface area and small volume should be useful in cases where it is desirable to minimize the volume of extracorporeal blood, or where the biological material is expensive (e.g. enzymes), or where a portable or wearable device is wanted.

Sparks' group (1969, 1971) extended this approach by suggesting the feasibility of artificial cells containing urease and ammonia absorbent or urea absorbent in the gastrointestinal tract or in the dialysis fluid of artificial kidneys for the removal of urea. The use of urease has also been modified by Gordon *et al.* (1969) for the removal of urea from dialysance. Other workers (Falb *et al.,* 1968; Flinn and Cherry, 1970; Gardner, 1971; Sparks *et al.,* 1971) are investigating the removal of urea by artificial cells containing different systems: urease and ammonia absorbent; urease and another enzyme system; or urea absorbent.

Work in this laboratory suggests the possible gastrointestinal approach for the removal of urea (Chang and Poznansky, 1968a) (Chang and Loa, 1970). In the intestine, bacterial urease converts urea into ammonia which is reabsorbed. Therefore, there is little intestinal excretion of urea or ammonia. However, in the present experiments, four hours after the oral administration of zirconium phosphate (ammonia absorbent), the systemic blood urea levels were 11.0 ± 1.1 mg% as compared to 14.5 ± 1.5 mg% in the control rats. The serum calcium levels were 10.22 ± 0.77 mg% in the treated group and 10.65 ± 0.40 mg% in the control group. Similar results were observed when another ammonia absorbent, Dowex-50W-X12, was administered orally into rats. Within four hours, the systemic blood urea levels had decreased to 65.0 ± 4.5 percent of the control values. When animals were treated with an antibiotic mixture (sulfaguanidine, Terramycin®, penicillin, and neomycin) to eliminate the urease-producing bacteria, oral administration of ammonia absorbents had little effect on the systemic blood urea levels (89.1 ± 6.0 percent of the control values). In these antibiotic-treated rats, the oral administration of both microencapsulated urease and ammonia absorbents lowered the systemic blood urea levels to 60.6 ± 5.0 percent of the control levels within four hours. Thus microencapsulated urease efficiently replaced the bacterial urease activity. In conclusion, a combination of exogenous ammonia absorbents and bacterial or microencapsulated urease in the intestine can efficiently remove systemic urea. More recently, aldehyde starch has been tested for the removal of urea (Giordano and Esposito, 1970; Sparks et. al. 1971).

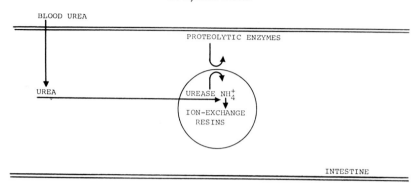

Figure 60. Schematic representation of the action of urease-loaded artificial cells in the gastrointestinal tract.

REMOVAL OF OTHER UREMIC METABOLITES

What can be placed inside the artificial cells to remove other uremic metabolites entering the artificial cells? While no ideal reactants are as yet available for this purpose, the enclosure of activated charcoal in artificial cells has been proposed (Chang, 1966) and studied in experimental animals (Chang et al., 1967b, 1968, 1970; Chang, 1969e), and in uremic patients (Chang and Malave, 1970; Chang et al., 1971).

It is known that activated charcoal can efficiently remove many uremic metabolites from perfusing blood and improve the symptoms of patients so treated (Yatzidis, 1964; Dunea and Kolff, 1965). Unfortunately, activated charcoal also adversely affects platelets of perfusing blood (Dunea and Kolff, 1965; Dutton et al., 1969) and releases embolising particles (Hagstam et al., 1966). On the other hand, when microencapsulated within semipermeable membranes (Chang, 1966), the activated charcoal is enclosed in an "intracellular environment." The enclosing membranes prevent any free powder from going into the circulation, and at the same time, prevent the blood platelets or plasma protein from coming into direct contact with the enclosed activated charcoal.

Compact Artificial Kidney System

The use of the standard procedure, as previously described (Chang, 1964; Chang et al., 1966), is not suitable for the microencapsulation of activated charcoal in a scaled-up system. The flexibility of the enclosed membranes results in packing and increased resistance to flow. The formation of artificial cells by polymer coating (Chang *et al.*, 1968; Chang, 1969e) prevents these problems and is the procedure which was used here. In the present study, coconut activated charcoal (6-14 mesh, Fisher Co.) is microencapsulated in albumin-coated collodion membranes (ACAC) by the procedures already reported (Chang, 1969e). After microencapsulation, instead of using the spray-drying procedure, the artificial cells are spread out and allowed to evaporate in a well ventilated 50°C oven until the ether has been removed (5 hours). They are then washed repeatedly with pyrogen-free distilled water through a sieve (40 mesh) until all free particles which have escaped microencapsulation are removed. In all experiments, and in the clinical trial, all artificial cells were used within 1 month after preparation.

Activated charcoal (300 gm) or artificial cells containing 300 gm of activated charcoal are placed in a silicone-coated (Silicone Spray, Manostat Co.) extracorporeal shunt chamber (10 cm diameter and 8 cm height) prepared as described previously (Chang, 1969e) (Fig. 61). The afferent shunt connection was prepared from sterile tubing (R62A Travenol), and the efferent connection, with air and clot traps, was prepared from a blood administration set (R78 Baxter). The priming volume of the connections and fully loaded shunt is about 300 ml.

Different types of sterilization procedures, including ethylene oxide, formalin, and autoclaving have been examined as possible sterilization methods for this system. Autoclaving at 121°C, 15 psi has been found to be suitable. Before microencapsulation, activated charcoal granules are washed to remove all fine particles. They are then autoclaved for 60 minutes. After further washing, they are wrapped and sealed in sterile cloth. The sealed portions are then autoclaved for 30 minutes, then allowed to dry completely at 60°C (at least 48 hours). After microencapsulation, evaporation, and the sieving procedure, the microencapsulated activated charcoal is placed in the shunt, primed with water, and autoclaved for 30 minutes. This process may disturb some artificial cell membranes, and careful perfusion at 200 ml/min with sterilized saline

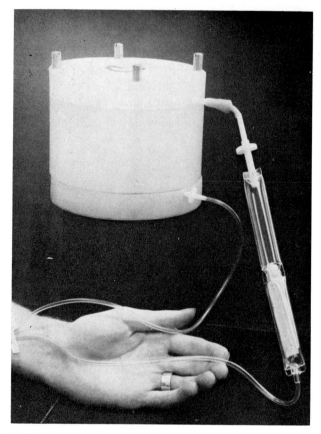

Figure 61. Extracorporeal shunt chamber containing 300 gm of ACAC artificial cells used for *in vivo* experiments and for clinical trials. (From Chang
et al., 1970. Courtesy of the American Society for Artificial Internal Organs.)

(4 liters) is required to remove any possible free particles. This is
followed by albumin coating; using sterilized technique, the shunt
is filled with 1 gm% human albumin (Cutter Co.), then left
standing at 4°C for 12 hours. Just before use, the shunt is perfused
with 2 liters of sterilized saline. Heparin (1000 units) is injected
into the shunt and 300 ml of saline is displaced by the patient's
blood before the effluent is returned to the patient. The patient
is heparinized systemically with 100-120 units/kg of heparin. At
present, a 200 ml/min blood flow can be obtained without the

use of a blood pump if the Scribner type of arteriovenous shunt is available. For the internal A-V fistuli, a blood pump is required. In these cases, a rotary pump and not a Sigma pump should be used.

The internal resistance of 35 mm Hg across the ACAC artificial cell type artificial kidney is low when compared to those of the conventional artificial kidneys: 100-300 mm Hg for the coil type artificial kidneys; 50 mm Hg for the Kiil type; and 45 mm Hg for the Dow type (Cestero and Freeman, 1969).

In Vitro and *In Vivio* Removal Of Uremic Metabolites

Creatinine. This was studied in detail in this report as a test of the efficiency of the artificial cell system. The *in vitro* removal of creatinine was first studied using stirred-batch experiments at 37°C. ACAC microencapsulated activated charcoal was added to a well-stirred aqueous creatinine solution (21 mg%) in amounts of 1 gm/30 ml of solution. The concentration of creatinine decreased at a rapid rate (Fig. 62); 50 percent of the creatinine was removed within 12 minutes.

Next, 300 gm of ACAC microencapsulated activated charcoal was used in single pass studies in which the creatinine concentration entering the shunt was kept at 20 mg%. Figure 63 shows the flow rates and time of perfusion. Creatinine clearance was calculated from the difference in creatinine concentration entering and leaving the shunt and the shunt flow rate. The results obtained are shown in Figure 63 and compared to other types of artificial kidneys in Figure 70. These studies show a higher creatinine clearance for 300 gm of ACAC microencapsulated activated charcoal *in vitro* than for the best conventional artificial kidneys (Fig. 70).

Further *in vitro* studies were done using a bath containing 42 liters of creatinine solution (21 mg%) (Fig. 64). Bath solution was recirculated continuously through a shunt containing 300 gm of ACAC microencapsulated activated charcoal. Within two hours the level of creatinine in the bath solution was lowered by 25 percent, representing the removal of 2.2 gm of creatinine within this period of time.

In vivo experiments in 20 bilaterally nephrectomized dogs were

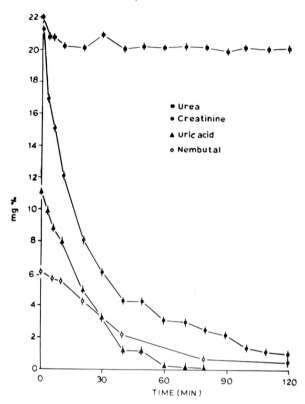

Figure 62. Rate of decrease in concentration of different solutes in stirred batch experiments (equivalent to 300 gm ACAC artificial cells in 9 liters of solution). (From Chang and Malave, 1970. Courtesy of the American Society for Artificial Internal Organs.)

done (Fig. 65). The experimental procedures described earlier (Chang, 1969e) were used. Briefly, arterial blood from each Nembutal-anesthetized dog (20-27 kg) was perfused through a shunt containing 300 gm of microencapsulated activated charcoal. A flow rate of 120 ml/min was possible without the use of blood pumps. Figure 65 shows the results obtained in 10 dogs treated with 300 gm of ACAC (albumin-coated collodion microencapsulated activated charcoal). In all cases (Fig. 66) the decrease in arterial creatinine level after two hours of hemoperfusion was about 35 percent; with a higher blood flow of 300 ml/min, the

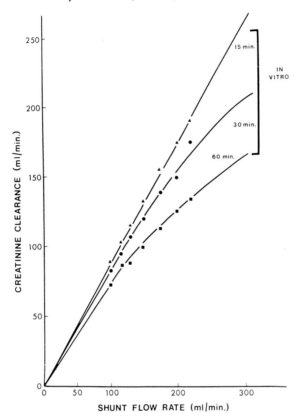

Figure 63. *In vitro* creatinine clearance of a shunt containing 300 gm of ACAC artificial cells. (From Chang and Malave, 1970. Courtesy of the American Society for Artificial Internal Organs.)

decrease was 75 percent. The decrease was most rapid in the first 30 minutes of hemoperfusion. Figure 65 compares the results obtained using albumin-coated collodion microencapsulated activated charcoal (ACAC) with those using free activated charcoal (AC) and those using collodion microencapsulated charcoal (CAC). ACAC and CAC microencapsulated activated charcoal removed plasma creatinine at a rate which was only slightly slower than free activated charcoal.

Uric acid. Figure 62 shows the result of stirred-batch experiments. Uric acid was removed efficiently by ACAC microencap-

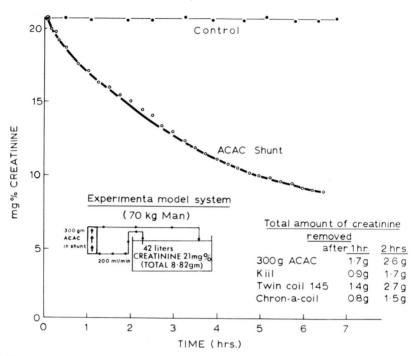

Figure 64. Schematic representation of the removal of creatinine by 300 gm of ACAC artificial cells.

sulated activated charcoal *in vitro*. Thus, 300 gm of this material removed 50 percent of the uric acid in 9 liters of an 11 mg% solution within 18 minutes. *In vivo* experiments in five dogs also showed that uric acid was removed efficiently.

Urea, Ca^{++}, phosphate, albumin, and total protein. As shown in Figure 62, ACAC microencapsulated activated charcoal only removed a very small amount of urea. *In vivo* experiments showed, as in the case of free activated charcoal, that ACAC microencapsulated activated charcoal had little effect on the removal of urea, Ca^{++}, phosphate, plasma albumin, or total protein.

Guanidine. Guanidines are efficiently removed by artificial cells containing activated charcoal.

Clinical Tests For Safety Of The System

Experiments described in the last chapter showed that albumin-

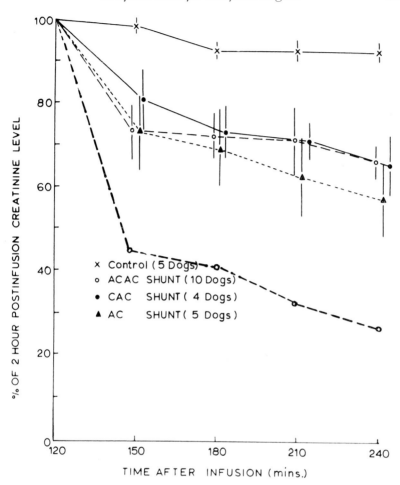

Figure 65. Rate of change of plasma creatinine levels in nephrectomized dogs treated with hemoperfusion (blood flow 120 ml/min) through free AC activated charcoal, CAC artificial cells, and ACAC artificial cells; 300 gm in each shunt. Lower curve (o) represents rate of change of plasma creatinine with higher blood flow (300 ml/min) through 300 gm of ACAC artificial cells. (From Chang and Malave, 1970. Courtesy of American Society for Artificial Internal Organs.)

coated artificial cells containing activated charcoal did not have any adverse effects on the formed elements of blood; no emboli were observed, and animals continued to do well in long-term

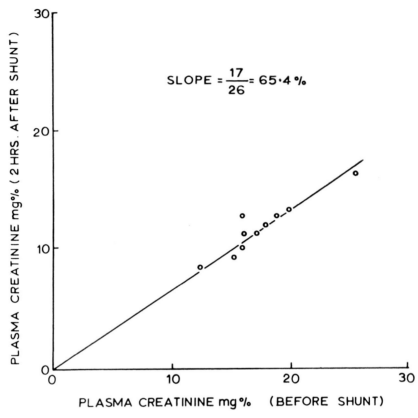

Figure 66. Effects of two-hour hemoperfusion through 300 gm of ACAC artificial cells on plasma creatinine levels of 10 nephrectomized dogs. (From Chang and Malave, 1970. Courtesy of the American Society for Artificial Internal Organs.)

studies. Thus at the Royal Victoria Hospital we initiated clinical trials of this system for chronic renal failure (Chang *et al.*, 1970). The first clinical trials were to test the safety of the system, and subsequent trials were to test the efficiency of the system. The first clinical test is summarized below (Fig. 67):

> B.B. is a 50-year-old white male with chronic renal failure and chronic lung disease. The patient could not be accommodated in either a chronic hemodialysis or a renal transplantation program and had been maintained by peritoneal dialysis. On the day of the clinical

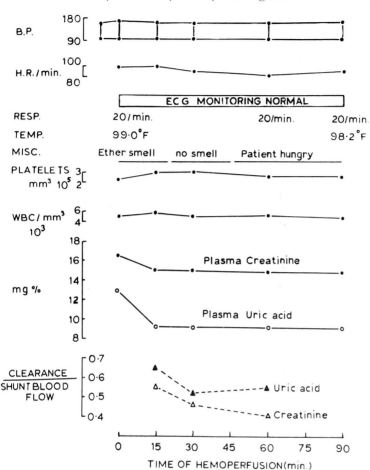

Figure 67. Response of patient (B.B.) treated with hemoperfusion through 300 gm of ACAC artificial cells. (From Chang and Malave, 1970. Courtesy of the American Society for Artificial Internal Organs.)

trial, his biochemical data were as follows: BUN 102 mg%, creatinine 16.5 mg%, uric acid 12.9 mg%, sodium 143 mEq/litre, chloride 98 mEq/litre, potassium 4.4 mEq/litre, calcium 7.4 mg%, and phosphorus 6.4 mg%.

The extracorporeal shunt chamber containing 300 gm of albumin-coated collodion microencapsulated activated charcoal (ACAC) was used (Fig. 61). Sterilization of the system was as described. After percutaneous puncture of the left femoral artery (14-gauge catheter),

3000 U.S.P. units of heparin was injected into the shunt and 5000 U.S.P. units intravenously into the patient. Blood from the left femoral artery entered the shunt to displace the 350 ml saline priming solution. After discarding the 350 ml priming solution, the effluent flow from the shunt was returned to the patient by the cephalic vein (14-gauge catheter). No blood pump was used, and a shunt blood flow of 100 ml/min was maintained.

At the beginning of the hemoperfusion procedure, the patient stated that there was a smell of ether. It was found that this came from a trace amount of residual ether present in the collodion membranes. This smell disappeared shortly. The patient felt well throughout the 90-minute procedure. No other side effects were noted. Instead of nausea and vomiting encountered by some patients treated with free activated charcoal hemoperfusion (Dunea and Kolff, 1965) our patient started to feel hungry halfway through the procedure. The patient accepted this procedure so completely that immediately after the procedure, he agreed to have repeated hemoperfusion in the future. The patient was followed for 48 hours with no adverse effects and was discharged from the hospital. He is scheduled for further hemoperfusions. Since then, the patient has received another hemoperfusion procedure with no adverse effects.

Hematological data from this laboratory showed that 90 minutes of hemoperfusion did not result in any adverse effects on platelets (237,000/mm³ initial level to 234,000/mm³ after hemoperfusion); neither leucocytosis nor leucopenia was noted (5,170/mm³ initial level to 4,290/mm³ after hemofusion); and there was no significant increase in the plasma hemoglobin level.

Performance Characteristics in Clinical Trials

The results obtained in 51 clinical tests (Chang *et al.*, 1971) are summarized below:

Internal resistant. The internal resistance measured as pressure drop across the microcapsule artificial kidney is very low (Fig. 73). At a blood flow rate of 200 ml/min the internal resistance of the artificial kidney prepared from the 6-10 mesh microcapsules (25 mm Hg) is less than that prepared from the 6-14 mesh microcapsules (45 mm Hg). In both cases, the internal resistances are less than the Kiil-type hemodialyzer (50 mm Hg). This internal resistance is extremely low when compared to the coil-type hemodialyzer (mostly over 150 mm Hg). Thus, in patients with Scribner A-V shunts, a blood flow rate of 150-260 ml/min can be obtained with the microcapsule artificial kidneys

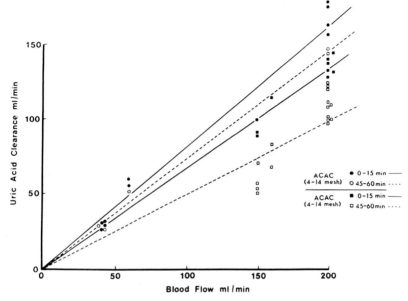

Figure 68. Creatinine clearance of 300 gm of albumin-collodion coated activated charcoal (ACAC) in 15 clinical trials. (From Chang *et al.*, 1971. Courtesy of the American Society for Artificial Internal Organs.)

without the use of a blood pump. In those patients with internal A-V fistuli, blood flow was much slower.

Creatinine. In vitro studies show that at a flow rate of 200 ml/min, 300 gm of albumin-coated microencapsulated activated charcoal can remove 5 gm of creatinine. The result of the removal of creatinine in the 51 clinical trials is summarized in Figures 68 and 70. It is clear that within a two-hour hemoperfusion period, the creatinine clearance of 300 gm of albumin-coated microencapsulated activated charcoal is comparable to the most efficient types of coil artificial kidney and much more efficient than the plate-type or capillary-type artificial kidneys.

Uric acid. The uric acid clearance in the 51 clinical trials is shown in Figures 69 and 71. It is clear that uric acid clearance is even more efficient than the standard hemodialyzers.

Guanidine. In vitro studies show that guanidine can be removed efficiently by microencapsulated activated charcoal. Pre- and post-hemoperfusion values for guanidine were measured in four

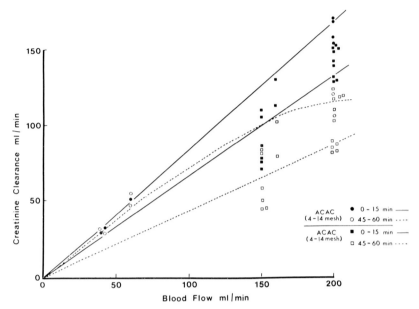

Figure 69. Uric acid clearance of 300 gm of albumin-collodion coated activated charcoal (ACAC) in 15 clinical trials. (From Chang *et al.*, 1971. Courtesy of the American Society for Artificial Internal Organs.)

clinical trials. In one of these, the analysis was interfered by a very high blood urea level. The results of the three successful analyses are shown in Table V. It is shown that guanidine, an important uremic metabolite, can be efficiently removed.

Urea, water, and electrolytes. The 300 gm of microencapsulated activated charcoal only removed a small fraction of blood urea. Other systems are being developed for the removal of urea (Chang, 1966; Levine and LaCourse, 1967; Chang and Loa, 1970; Falb *et al.*, 1970; Sparks *et al.*, 1971; Gardner, 1971). Microencapsulation of possible urea absorbents is also being investigated (Sparks *et al.*, 1971). The present microencapsulated activated charcoal system does not contribute to the removal of sodium, potassium, phosphate, or water. Microncapsulation of ion-exchange resins (Chang, 1966) is being investigated.

Effects on platelets and safety of system. The posthemoperfusion level in the case of albumin-coated microencapsulated acti-

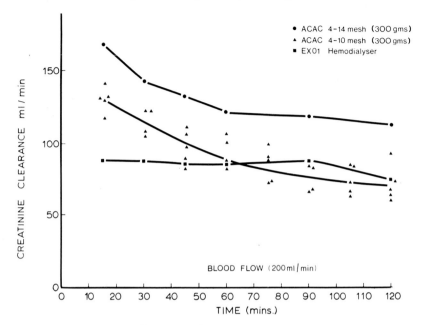

Figure 70. Creatinine clearance of 300 gm of albumin-collodion coated activated charcoal (ACAC) in clinical trials compared to those obtained using EX01 hemodialyzers. Blood flow rate 200 ml/min. (From Chang *et al.*, 1971. Courtesy of the American Society for Artificial Internal Organs.)

vated charcoal was 91.8%±11.8% of control values (Fig. 72). This is a negligible change when compared to the 50% post-hemoperfusion level in uncoated activated charcoal (Dunea and Kolff, 1965; DeMyttenaere *et al.*, 1967; Hagstam *et al.*, 1966; Dutton *et al.*, 1969) and the 70% posthemodialysis level in coil hemodialyzer (Lawson *et al.*, 1966). Smear of shunt effluent blood did not show any embolic particles. These observations in clinical trials confirm our earlier findings that hemoperfusion over albumin-coated microencapsulated activated charcoal did not give rise to embolising particles or changes in platelet levels (Chang, 1969; Chang and Malave, 1970).

Transient pyrogenic reactions were observed in two of the earlier hemoperfusions. The blood cultures were all negative. Pyrogenic tests showed that the albumin-coated microencapsulated activated charcoal was not pyrogenic. The most likely

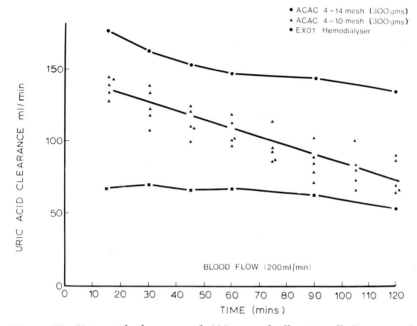

Figure 71. Uric acid clearance of 300 gm of albumin-collodion coated activated charcoal (ACAC) in clinical trials compared to those obtained using EX01 hemodialyzers. Blood flow rate 200 ml/min. (From Chang *et al.*, 1971. Courtesy of the American Society for Artificial Internal Organs.)

cause may have been due to the reuseable screens of the shunt chambers. Since then, careful washing of the reusable shunt chambers with sodium hypochlorite to remove any protein deposited in the wire meshes and elsewhere has eliminated the problems of pyrogenic reactions. No other side effects were observed.

Chronic Intermittent Hemoperfusions with the Microcapsule Artificial Kidney

The microcapsule artificial kidney is at present being used for maintainance chronic hemoperfusion (Chang *et al.*, 1971). The patient is a 71-year-old Italian woman with chronic renal failure and congestive heart failure. Her general condition was such that she was not suitable for either the transplantation or chronic

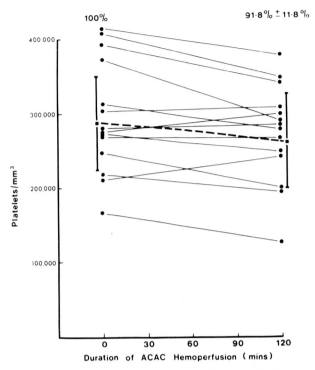

Figure 72. Effects of hemoperfusion over albumin-collodion coated activated charcoal on patients' arterial platelet levels. (From Chang *et al.*, 1971. Courtesy of the American Society for Artificial Internal Organs.)

hemodialysis program. After peritoneal dialysis, she was placed on medication and diet. The patient continued with her medication and diet but did not return for her subsequent peritoneal dialysis. She was admitted to the hospital three months later with complaints of nausea, vomiting, low back pain, diarrhea, and hiccups. Her biochemical data at the time of admission was as follows: BUN 186 mg%, creatinine 24.9 mg%, Na 142 mEq/litre, K 5.8 mEq/litre, and Cl 102 mEq/litre. Her urinary output was about 600 ml/24 hours and creatinine excretion was about 80 mg/24 hours. She was placed on a regime of intermittent microcapsule hemoperfusions supplemented with hemodialysis as required. In addition, she continued her medication of Gravol, Amphojel®, folic acid, Kayexalate®, sodium bicarbonate,

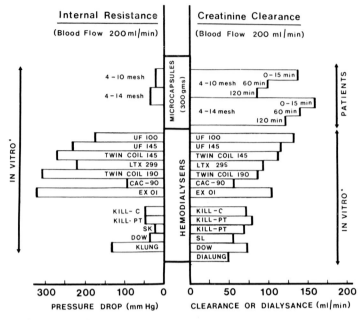

Figure 73. Internal resistance and creatinine clearance of microcapsule (ACAC) artificial kidney compared to the values obtained by Cestero *et al.* for other hemodialyzer. (From Chang *et al.*, 1971. Courtesy of the American Society for Artificial Internal Organs.)

TABLE V

REMOVAL OF GUANIDINE

Microcapsules (GMS)	Blood Flow (ml/min)	Duration (hours)	Arterial Guanipine Level	
			Prehemoperfusion (mg%)	Posthemoperfusion (mg%)
300	200	2	1.0	0.5
300	150	2	1.6	0.4
600	150	3	2.2	0.8

and Bemianl® with C fortis; and her restriction to 40 gm protein, 40 mq Na and 40 mq K. The hemoperfusion and hemodialysis schedule and the biochemical data is shown in Figure 74. Throughout these hemoperfusions, the patient did not experience any pyrogenic reactions or other side effects. During two of the 28 hemoperfusions, there was a slight decrease in her blood pressure (130/80 to 110/70) during the procedure. However, her blood pressure immediately returned to her preperfusion levels at the completion of the two-hour procedure when the extracorporeal blood in the chamber was reinfused into the patient. As shown in Figure 72, the changes in posthemoperfusion platelet levels are negligible. Furthermore, her systemic platelet levels show a steady increase after the initiation of this regime, (Fig. 74). There is no hemolysis or changes in her plasma hemoglobin levels. Her urinary output remains at 200-600 ml/24 hours. She is doing well after 28 treatments in 8 months.

The creatinine and uric acid clearance obtained are shown in Figures 68, 69, 70, and 71, along with the other clinical tests, and compared with the standard hemodialysis in Figure 73. Both systemic creatinine and uric acid levels fell after each hemoperfusion (Fig. 74). The decrease depends on blood flow rate and amount of microcapsules used and ranged from 3.0 mg% to 5.0 mg% in two hours with a hemoperfsuion rate of 150 ml/min to 200 ml/min over 300 gm of microcapsules. In the one case where two 300 gm microcapsules shunts were used one after another at a hemoperfusion rate of 150 ml/min, there was a decrease of about 7.0 mg% within three hours. This rate of decrease of systemic creatinine level is more efficient than that of the EX01 hemodialyzer when calculated for the same blood flow rate. Her arterial blood guanidine levels were successfully measured in two of the clinical tests. With two hours of hemoperfusion at 150 ml/min over 300 gm of microcapsules, the arterial level fell from 1.6 mg% to a posthemoperfusion level of 0.4%. With three hours of hemoperfusion at 150 ml/min over 600 gm of microcapsules, the arterial level fell from 2.2 mg% to a posthemoperfusion level of 0.8 mg%. While discussing the removal of guanidine, it is interesting to note that with the present regime of intermittent hemoperfusion, medications and diet, the patient has remained relatively asymptomatic for the last 8 months. She is more active and her appetite has increased and she only returns to the hospital for her hemoperfusion and 2 hours posthemoperfusion observations. Her problem is water retention and ascite. More detailed studies will be carried out in this and other patients

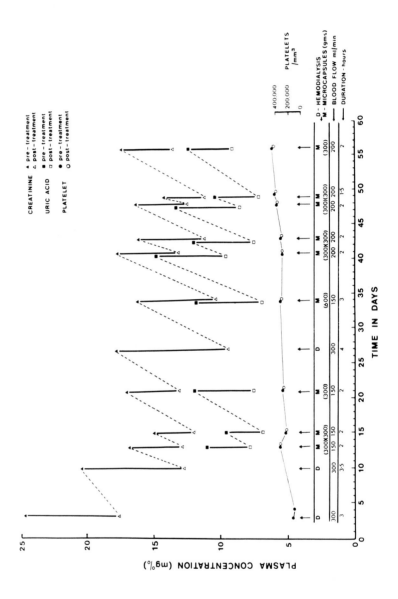

Figure 74. Patient treated with chronic intermittent hemoperfusion with microcapsules artificial kidney over a two months period. (From Chang *et. al.*, 1971. Courtesy of the American Society for Artificial Internal Organs.)

to establish the optimal frequency and duration of intermittent microcapsule hemoperfusions and the precise effect of hemoperfusion on the patients' symptomatology.

DISCUSSION

Detailed *in vitro, in vivo* and clinical results indicate that activated charcoal enclosed in artificial cells is prevented from adversely affecting the formed elements of blood and from giving off massive emboli. The clearance of creatinine and uric acid by 300 gm of ACAC artificial cells is much more efficient than the conventional artificial kidneys (Fig. 73). The compactness, the absence of a large volume of dialysate fluid, and the ease of operation all favor the suggestion of using artificial cells for the construction of compact artificial kidneys (Chang, 1966). There are other substances such as urea, electrolytes, and water which require removal. Work on the possible uses of microencapsulated urease absorbents or of urease and ammonia absorbents for the removal of urea is in progress (Fig. 75) (Chang, 1966, 1970; Gardner *et al.*, 1971; Sparks *et al.*, 1969). Micorencapsulated ion exchange resins have been prepared (Chang, 1966); selection of the appropriate resins may allow the removal of selected electrolytes. Even before these can be developed for clinical use, the microencapsulated activated charcoal system

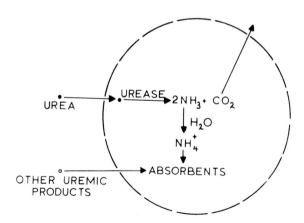

Figure 75. Schematic representation of an artificial cell containing urease and absorbents. (From Chang, 1969. Courtesy of Science Tools, Sweden.)

can be supplemented by less frequent conventional dialysis. In addition, the microencapsulated activated charcoal can be a simple and useful means of treating acute intoxication.

It should be emphasized that even though the present ACAC artificial cell system is compact, theoroeically it can be made much smaller, e.g. 10 ml (Chang, 1966) or 30 ml (Chang, 1970). Attaining this miniature size, however, would depend on the development of an ideal high capacity absorbent, or discovery of the unknown uremic toxin(s) which can then be selectively removed. The present system requires heparinization. The development of a nonthrombogenic system would greatly simplify the whole procedure. Work on the use of nonthrombogenic artificial cells is already in progress (Chang *et al.*, 1967b, 1968; Chang, 1969e) with promising results. The ultrathin membranes of the ACAC artificial cells withstand gravitational flow, blood flow, and rotatary blood pump flow of 200 ml/min or more, but vigorous pulsatile pumping with the sigma blood pump over long periods can affect the membranes.

The general idea of artificial cells (Chang, 1964, 1966) should not be confined to the construction of artificial kidneys. The present system is a feasibility test to study the use of artificial cells for the design of artificial organs. As such, the results obtained encourage one to proceed further. In nature, cells are the basic unit structures of all organs and tissues. It is left to one's imagination to extend this further into their possible uses to support hepatic failure and other metabolic disturbances. In fact, experiments have shown the possible application of this principle both in enzyme replacement for enzyme deficiency diseases (Chang and Poznansky, 1968a; Chang, 1969d) and in the treatment of asparagine-dependent lymphosarcoma (Chang, 1969c). It is hoped that the basic idea of "artificial cells" will be explored and developed further for the construction of artificial organs. The combination of the capillary dialysis system and the artificial cells being studied in this laboratory is a further step towards this goal.

Chapter 10

GENERAL DISCUSSIONS AND PERSPECTIVES

T HE biomedical implications of artificial cells have been investigated in this laboratory for more than 15 years. Since then, an increasing number of other laboratores have entered into this area of research, thus greatly extending and increasing the scope and the approach to the general idea of artificial cells. However, it is clear that although progress has been made, the whole area of research on artificial cells is only in its embryonic stage.

There is much work to be done in the development of artificial cell membranes. One approach would involve the complete reconstitution of biological cell membranes including the incorporation of the various transport mechanisms. Although some studies have been initiated to incorporate proteins, mucopolysaccharides, lipids, and transport systems into the artificial cell membranes, further progress depends greatly on the availability of further basic knowledge. Even though basic research in biological membranes is advancing rapidly, insufficient information is available at present to allow one to prepare artificial cells having completely reconstituted biological membranes. A more immediate and fruitful approach is the development of biologically compatible artificial cell membranes having characteristics suitable for the particular functions for which the artificial cells are to be used. A typical example is the development of nonthrombogenic artificial cells for use in extracorporeal perfusion. In other cases, where artificial cells are to be injected, suitable membranes will have to be developed so that the artificial cells can remain at the sites of injection, circulate in the bloodstream, or arrive at predetermined final sites. Membrane composition, porosity, or charge can be further modified so that the permeability characteristics can be more selective. Transport mechanisms might be incorporated into the artificial cell membranes to facilitate the even more selective movements of certain molecules. Until a few years ago, the vast amount of industrial microencapsulation technology had not been used for the preparation of the type of artificial cells described in this monography. Continued effort in this area should result in greater versatility in the preparation of artificial cells.

In addition to the unlimited possible variations in membrane properties, an even greater number of materials can be selected for enclosure in the artificial cells. This monograph only touches on a minute fraction of these possibilities. Even if we are to confine ourselves to the examples given in this monograph, much remains to be extended.

Thus, further work needs to be done to prepare artificial cells containing cell homogenates which could be administered by injection for the replacement or supplement of deficient body cells. Erythrocyte hemolysate has been enclosed in artificial cells. The major components, hemoglobin, carbonic anhydrase, and catalase, continued to function in the artificial cells. These results encourage the hope that further progress can be made in this direction, especially to enclose homogenate or cells of endocrine or liver tissues. Simplified versions of artificial cells containing one or more enzyme systems should also be explored. The demonstrated efficiency or artificial cells containing catalase for the experimental enzyme replacement of a "molecular disease," acatalasemia, suggests that further work should now be initiated to apply this system to other molecular diseases in clinical situations where such therapy might benefit the patients. A number of problems will no doubt have to be solved before this basic finding can be put to clinical uses. First, cell homogenates or suitable enzyme systems will need to be extracted or synthesized so that they can be loaded into artificial cells. In some other cases, where cofactors are required, one may have to include the cofactor with the enzyme or include enzyme systems which can generate the cofactor. Continuation of studies along these lines may contribute to the fulfillment of Pauling's 1956 prediction that by the year 2005, clinical enzyme replacement therapy for enzyme deficiency will become a reality. The results obtained using asparaginase-loaded artificial cells for the experimental suppression of mice lymphosarcoma encourage further work to investigate the use of this system for the treatment of substrate-dependent tumours. In addition, it would be of interest to see whether changes in the levels of various plasma constitutents could be produced by means of artificial cells containing other enzyme systems. Examples of substrate-enzyme pairs that might be studied in this way would be uric acid and uricase, choline and choline oxidase, dopa (diphydroxyphenylalanine) and dopa decarboxylase, histidine and its decarboxylase, histamine and diamine oxidase, glucose and glucose oxidase, angiotensin and angiotensinase. Finally, it would be of interest to test the possibility of using enzyme-loaded artificial cells to combat experimental intoxication for instance acetylcholinesterase against organic phosphate insecticides, or aldehyde dehydrogenase against methanol poisoning.

The uses of materials other than those normally present in the biological cells for the preparation of biomedically useful artificial cells have much promise. Thus the uses of artificial cells containing detoxicants for the removal of exogenous and endogenous toxins have already been demonstrated. This system should be extended and improved for clinical uses. Further modifications along these lines using other synthetic materials for other purposes should yield fruitful results. Thus, the use of synthetic materials like silicones or fluorocarbons for the prepartaion of red blood cell substitutes is already proving to be extremely exciting. Artificial cells containing both biological and synthetic materials is another area which should be explored. A typical example is the combination of enzymes and synthetic materials which can remove or modify the product of the enzymatic reaction.

Since cells are the unit structures of all organs and tissues, the possible uses of artificial cells as basic units for the construction of artificial organs deserve intensive investigation. This is especially so since experimental and preliminary clinical data have already demonstrated the feasibility of using the principle of artificial cells for the construction of compact artificial kidneys. Other organs, especially those concerned with metabolism, should be good candidates for this type of investigation.

As emphasized at the beginning of this monograph, it will be some time before we can achieve the complete reconstitution of biological cells. Nevertheless, even before this ultimate goal can be reached, the principle of artificial cells has already been shown to have a number of clinical implications. "Artificial cell" is a concept; the examples described in this monograph are but physical examples for demonstrating this idea. In addition to extending and modifying the present physical examples, completely different systems could be made available to further demonstrate the clinical implications of the idea of "artificial cells." Individual research investigators working independently can greatly extend the frontiers of this area of research. On the other hand, the obvious interdisciplinary nature of this research is such that it will require a closely coordinated biomedical group effort, consisting of chemists, chemical engineers, biophysicists, basic medical scientists, and clinicians, to develop these basic findings towards early clinical applications. The potential of artificial cells in biomedical research and clinical applications is limited only by one's imagination. An entirely new horizon is waiting impatiently to be explored.

REFERENCES

Able, J. J., Rowntree, L. G., and Turner, B. B. (1913): On the removal of diffusible substances from circulating blood by means of dialysis. *Trans. Ass. Amer. Physicians, 28:* 51.

Abramson, H. A. (1934): *Electrophoretic Phenomena.* Chemical Catalog Co., New York.

Adamson, R. H., and Fabro, S. (1968): Antitumor activity and other biologic properties of L-asparaginase. *Cancer Chemother. Rep., 52:*617.

Algire, G. H., Weaver, J. M., and Prehn, R. T. (1954): Growth of cells *in vivo* in diffusion chambers: survival of homografts in immunized mice. *J. Nat. Cancer Inst., 15:*493.

Alwall, N. (1947): On the artificial kidney; I. Apparatus for dialysis of blood *in vivo. Acta. Med. Scand., 128:*317.

Anderson, N. (1967): Light transmission and scattering properties of flowing suspensions with particular reference to red blood cells. Ph.D. Thesis, McGill University, Montreal.

Andjus, R., Suhara, K., and Sloviter, H.A. (1967): An isolated perfused rat brain preparation, its spontaneous and stimulated activity. *J. Appl. Physiol., 22:*1033.

Andrade, J. D., Kunitomo, K., Lim, D., and Kloff, W. J. (1971a): Coated activated carbon as an atraumatic adsorbent for direct blood perfusion. *Fed. Proc., 30:*705 (Abs).

Andrade, J. D., Kunitomo, K., Wagenon, R., Gough, D., and Kolff, W. (1971b): Coated adsorbents for direct blood perfusion: HEMA activated carbon. *Trans. Amer. Soc. Artif. Organs)* (in press).

Axen, R., Porath, J., and Ernback, S. (1967): Chemical coupling of peptides and proteins to polysaccharides by means of cyanogen halides. *Nature, 214:*1302.

Babb, A. L., Popovich, R. P., Christopher, T. G., and Scribner, B. H. (1971): The genesis of the square meter-hour hypothesis. *Trans. Amer. Soc. Artif. Intern. Organs* (in press).

Bakan, J., and Anderson, J. (1970): Microencapsulation. In Kanig, J. L., Lachman, L., and Lieberman, H. A. (Eds.): *Industrial Pharmacy,* Lea & Febiger, Philadelphia.

Bangham, A. D., Standish, M. M., and Watkin, J. A. (1965): Diffusion of univalent ions across the lamellae of swollen phosphalipids. *J. Molec. Biol., 13:*238.

Basset, C., Andrew, C., and Campbell, J. B. (1960): Calcification of millipore *in vivo. Transplant. Bull., 26:*132.

Berlin, N. I., Walsmann, T. A., and Weissman, S. M. (1959): Membrane transport. *Physiol. Rev.*, 93:577.

Bock, J. C. (1920): A study of a decolorizing carbon. *J. Amer. Chem. Soc.*, 42:1564.

Booig, H. L., and Bungenberg de Jong, H. F. (1956): Bicolloids and their interactions. *Protoplasma, 1*:2.

Brest, A. N., and Moyer, J. H. (1967): *Renal Failure.* Lippincott, Philadelphia.

Braley, S. A. (1960): *Bulletin of Dow Corning Centre for aid to Medical Research, 2*:8.

Brooks, J. R., and Hill, G. H. (1960a): A study of the survival and function of homografted thyroid tissue in membrane filter chambers in the rat. *Endocrinology, 66*:393.

Brooks, J. R., Sturgis, S. H., and Hill, G. J. (1960b): An evaluation of endocrine tissue homotransplantation in the millipore chamber; with a note on tissue adaptation to the host. *Ann. N.Y. Acad. Sci., 87*:482.

Broome, J. D. (1961): Evidence that the L-asparaginase activity of guinea pig serum is responsible for its antilymphoma effects. *Nature, 191*:1114.

Broome, J. D. (1963): Evidence that the L-asparaginase of guinea pig serum is responsible for its antilymphoma effects; I. Properties of the L-asparaginase of guinea pig serum in relation to those of the antilymphoma substance. *J. Exp. Med., 118*:99.

Broome, J. D. (1968): L-asparaginase: the evolution of a new tumor inhibitory agent. *Trans. N.Y. Acad. Sci., 30*:690.

Bungenberg de Jong, H. G., and Bonner, J. (1935): Phosphatide autocomplex coacervates as ionic systems and their relationship to the protoplasmic membrane. *Protoplasma, 24*:198.

Burgen, A. S. V. (1957): The physiological ultrastructure of cell membranes. *Canad. J. Biochem. Physiol., 35*:569.

Burton, B. T. (1967): *Kidney Disease Program Analysis.* U.S. Dept. of Health, Education and Welfare.

Canham, P. B., and Burton, A. C. (1968): Distribution of size and shape in populations of normal human red cells. *Cir. Res., 22*:405.

Carr, C. W., Gregor, H. P., and Sollner, K. (1945): The preparation and properties of "Megapermselective" protamine collodion membranes combining high ionic selectivity with high permeability. *J. Gen. Physiol., 28*:179

Cestero, R. V. M., and Freeman, R. B. (1969): Comparative performance characteristics of 13 hemodialysers. *Trans. Amer. Soc. Artif. Intern. Organs, 15*:81.

Chang, T. M. S. (1957): Hemoglobin corpuscles. Report of research project for B.Sc. Honours Physiology, McGill University, Montreal.

Chang, T. M. S. (1964): Semipermeable microcapsules. *Science, 146*:524.

Chang, T. M. S. (1965): Semipermeable aqueous microcapsules. Ph.D. Thesis, Physiology, McGill University, Montreal.

Chang, T. M. S. (1966): Semipermeable aqueous microcapsules ("artificial cells"): with emphasis on experiments in an extracorporeal shunt system. *Trans. Amer. Soc. Artif. Intern. Organs*, *12*:13.

Chang, T. M. S. (1967): Microcapsules as artificial cells. *Science J.*, *3*:63.

Chang, T. M. S. (1969a): Artificial cells made to order. *New Scientist*, *42*:18.

Chang, T. M. S. (1969b): Lipid-coated spherical ultrathin membranes of polymer or cross-linked protein as possible cell membrane models. *Fed. Proc.*, *28*:461.

Chang, T. M. S. (1969c): Asparaginase-loaded semipermeable microcapsules for mouse lymphoma. *Proc. Canad. Fed. Biol. Sci.*, *12*:62.

Chang, T. M. S. (1969d): Clinical potential of enzyme technology. *Science Tools*, *16*:33.

Chang, T. M. S. (1969e): Removal of endogenous and exogenous toxins by a microencapsulated absorbent. *Canad. J. Physiol. Pharmacol.*, *47*: 1043.

Chang, T. M. S. (1969f): Discussion. *Trans Amer. Soc. Artif. Intern. Organs*, *15*:359.

Chang, T. M. S. (1970a): A new concept in artificial kidneys using semipermeable microcapsules. *Chemeca 70*, Australia Academy of Science and the Australian National Committee Institute of Chemical Engineers, Butterworth of Australia, p. 48.

Chang, T. M. S. (1971a): The *in vivo* effects of semipermeable microcapsules containing L-asparaginase on 6C3HED lymphosarcoma. *Nature*, *229*:117.

Chang, T. M. S. (1971b): Unpublished data.

Chang, T.M.S. (1972a): The effects of local applications of microencapsulated catalase in acatalasemia. J. Dent. Res. 2 Supp. 319.

Chang, T. M. S. (1971d): Stabilisation of enzymes by microencapsulation. Bioch. Biophy, Res. Comm. *44*:1531.

Chang, T. M. S., Gonda, A., Dirks, J. H., and Malave, N. 1971d): Clinical evalution of chronic intermittent or short term hemoperfusions in patients with chronic renal failure using semipermeable microcapsules (artificial cells) formed from membrane-coated activated charcoal. *Trans. Amer. Soc. Artif. Intern. Organs 17*, 246.

Chang, T. M. S., Johnson, L. J., and Ransome, O. (1967a): Semipermeable microcapsules with heparin-complexed membranes. *Proc. Canad. Fed. Biol. Sci.*, *10*:30.

Chang, T. M. S., Johnson, L. J., and Ransome, O. (1967b): Semipermeable aqueous microcapsules: Nonthrombogenic microcapsules with heparin-complexed membranes. *Canad. J. Physiol. Pharmacol.*, *45*:70.

Chang, T. M. S., Lesser, B., and Chuang, S. (1969): Unpublished data.

Chang, T. M. S., and Loa, S. K. (1970): Urea removal by urease and ammonia absorbent in the intestine. *Physiologist*, *13*:70.

Chang, T. M. S., MacIntosh, F. C., and Mason, S. G. (1963): Semipermeable aqueous microcapsules. *Proc. Canad. Fed. Biol. Sci.,* 6:16.

Chang, T. M. S., and MacIntosh, F. C. (1964a): Effects of injected semipermeable microcapsules. *Proc. Canad. Fed. Biol. Soc.,* 7:58.

Chang, T. M. S., and MacIntosh, F. C. (1964b): Semipermeable aqueous microcapsules. *Pharmacologist,* 6:198.

Chang, T. M. S., and MacIntosh, F. C. (1965): Semipermeable aqueous microcapsules. *Proceedings of the XXIII International Congress of Physiological Sciences,* Tokyo, Japan.

Chang, T. M. S., MacIntosh, F. C., and Mason, S. G. (1966): Semipermeable aqueous microcapsules: Preparation and properties. *Canad J. Physiol. Pharmacol.,* 44:115.

Chang, T. M. S., and Malave, N. (1970): The development and first clinical use of semipermeable microcapsules (artificial cells) as a compact artificial kidney. *Trans. Amer. Soc. Artif. Intern. Organs,* 16:141.

Chang, T. M. S., Pont, A., Johnson, L. J., and Malave, N. (1968): Response to intermittent extracorporeal perfusion through shunts containing semipermeable microcapsules. *Trans. Amer. Soc. Artif. Intern. Organs,* 15:163.

Chang, T. M. S., and Poznansky, M. J. (1968a): Semipermeable microcapsules containing catalase for enzyme replacement in acatalasemic mice. *Nature,* 218:243.

Chang, T. M. S., and Poznansky, M. J. (1968b): Microencapsulated catalase for acatalasemia. *Proc. Canad. Fed. Biol. Sci.,* 11:74.

Chang, T. M. S., and Poznansky, M. J. (1968c): Semipermeable aqueous microcapsules (artificial cells): permeability characteristics. *J. Biomed. Mat. Res.,* 2:187.

Chang, T. M. S., and Poznansky, M. J. (1968d): Catalase loaded semipermeable microcapsules for acatalesemia. *Proceedings of the XXIV International Congress of Physiological Sciences,* Washington, D.C.

Chang, T. M. S., and Rosenthal, A. (1969): Unpublished data.

Chemical Week (1965): What is new in microencapsulation. January 2, p. 45.

Chien, S. and Gregerson, M. I. (1962): Determination of body fluid volumes. In Nastuk, W. L. (Ed.), *Physical Techniques in Biological Research.* Academic Press, New York, vol. 4, p. 36.

Clark, L. C., Jr. (1967): Fluorocarbon emulsion as erythrocyte substitute. Edmund Hall Lecture, University of Louisville, Sigma Xi, May 19. W. Welch, *Louisville Times,* May 20.

Clark, L. C., Jr. (1970): Epilogue. *Fed. Proc.,* 29:1820.

Clark, L. C., Jr., and Gollan, F. (1966): Survival of mammals breathing organic liquids equilibrated with oxygen at atmospheric pressure. *Science,* 152:1755.

Clark, L. C., Jr., Kaplan, S., Becattini, F., and Benzing, G. III (1970): Perfusion of whole animal with perfluorinated liquid emulsions using the Clark bubble-defoam heart-lung machine. *Fed. Proc.,* 29:1794.

Cook, G. M. W., Heard, D. H., and Seaman, G. V. F. (1961): Sialic acids and the electrokinetic charge of the human erythrocyte. *Nature, 191:* 44.

Courtice, F. C. and Simmonds, W. J. (1954): Physiological significance of lymph drainage of serious cavities and lungs. *Physiol. Rev., 34:*419.

Craig, L. C. (1964): Differential dialysis. *Science, 144:*1093.

Danon, D., and Marikovsky, Y. (1961): Morphology of the membranes of young and old erythrocytes. Observations by electron microscope. *C. R. Soc. Biol., 155:*1271.

Davies, D. F. (1958): The electrophoretic mobility of erythrocytes as a measure of the surface activity of plasma from patients with and without evidence of atheroma. *Clin. Sci., 17:*563.

Davies, D. F., and Clark, A. (1961): An assessment of erythrocyte migration times and serum cholesterol levels as indices in myocardial infarction. *Clin. Sci., 20:*279.

Davson, H., and Danielli, J. F. (1943): *Permeability of Natural Membranes,* 1st ed. Harvard University Press, Cambridge, Mass.

Dean, H. M., Decker, C. L., and Baker, L. D. (1967): A patient with complete hemolysis of circulating red blood cell mass. *New Eng. J. Med., 277:*700.

Dedrick, R. L., and Bischoff, K. B. (1966): Chemical engineering in medicine. *Chem. Eng. Progr. Symp. Series, 66:*32.

Dedrick, R. L., Bischoff, K. B., and Leonard, E. F. (1968): The atrificial kidney. *Chem. Eng. Progr. Symp. Series, 84:*64.

DeMyttenaere, M. H., Maher, J. F., and Schreiner, G. E. (1967): Hemoperfusion through a charcoal column for glutethimide poisoning. *Trans. Amer. Soc., Artif. Intern. Organs, 13:*190.

Dobson, E. L. (1957): *Physiopathology of the Reticulo-endothelial System.* Blackwell, Oxford, p. 80.

Done, A. K. (1967): Treatment of salicylate poisoning. *Mod. Treatm., 4:*648.

Doyle, J. E. (1962): *Extracorporeal Hemodialysis Therapy in Blood Chemistry Disorders.* Thomas, Springfield.

Dunea, G. and Kolff, W. J. (1965): Clinical experiments with the Yatzidis charcoal artificial kidney. *Trans. Amer. Artif. Intern. Organs, 11:*178.

Dutton, R. C., Redrick, R. L., and Bull, B. S. (1969): A simple technique for the experimental production of acute platelet deficiency. *Thromb. Diath. Haemorrh., 21:*367.

Fairley, N. W. (1940): Renal clearance of hemoglobin. *Brit. Med. J. 2:* 213.

Falb, R. D., Anapakos, P. G., Nack, H., and Kim, B. C. (1968): Feasibility of a microcapsule system for artificial kidney application. *Annual Report,* Artificial Kidney Program, National Institute of Health.

Feinstein, R. N. (1970): Acatalasemia in the mouse and other species. *Bioch. Genetics, 4:*135.

Feinstein, R. N., Howard, J. B., Braun, J. T., and Seaholm, J. E. (1966a): Acatalasemic and hypocatalasemic mouse mutants. *Genetics, 53*:923.

Feinstein, R. N., Braun, J. T., and Howard, J. B. (1966b): Reversal of H_2O_2 toxicity in the acatalesemic mouse by catalase administration: suggested model for possible replacemental therapy of inborn error of metabolism. *J. Lab. Clin. Med., 68*:952.

Feinstein, R. N., Sutter, H., and Jaroslow, B. A. (1968): Blood catalase polymorphism: some immunological aspects. *Science, 159*:638.

Flinn, J. E., and Cherry, R. H., Jr. (1970): Separation technology. *Chem. Eng. Progr. Symp. Series, 65*:90.

Flinn, J. E., and Nack, H. (1967): What is happening in microencapsulation. *Chem. Eng., 171*:8.

Folkman, J., Cole, P., and Zimmerman, S. (1966): Tumor behavior in isolated perfused organs: *in vitro* growth and metastases of biopsy material in rabbit thyroid and canine intestinal segment. *Ann. Surg., 164*:491.

Folkman, J., and Long, D. M. (1963): Diffusion of drugs across silastic rubber. *Bulletin of the Dow Corning Centre for Aid to Medical Research, 5*:9.

Fourt, L., Schwartz, A. M., Quasius, A., and Bowman R. L. (1966): Heparin-bearing surfaces and liquid surfaces in relation to blood coagulation. *Trans. Amer. Soc. Artif. Intern. Organs, 12*:155.

French, E. L., and Ada, G. L. (1954): Action of receptor destroying enzyme of V. cholera (RDE) in guinea pigs. *Aust. J. Exp. Biol., 32*:165.

Gabourel, J. D. (1961): Cell culture *in vivo;* II. Behavior of L-fibroblasts in diffusion chambers in resistant hosts. *Cancer Res., 21*:506.

Gardner, D. L. (1971): Possible uremic detoxification via oral ingested microcapsules. *Trans. Amer. Soc. Artif. Intern. Organs* (in press).

Garrod, A. E. (1909): *Inborn Errors of Metabolism.* Henry Hrowde, Hodder & Stoughton, London.

Geyer, R. P. (1970): Whole animal perfusion with fluorocarbon dispersion. *Fed. Proc., 29*:1758-1763.

Geyer, R. P., Monroe, R. G., and Taylor, K. (1968): In Norman, J. C., et al. (Eds.), *Organ Perfusion and Preservation.* Appleton-Sentury-Crofts, New York.

Giordano, C. and Esposito, R. (1970) Oxystarch. Report to NIAMD, 1970.

Goldstein, D. A., and Solomon, A. K. (1960): Determination of equivalent pore radius for human red cells by osmotic pressure measurement. *J. Gen. Physiol., 44*:1.

Gollan, F., and Clark, L. C. (1966): Organ perfusion in fluorocarbon fluid. *Physiologist, 9*:191.

Gordon, A., Greenbaum, M. A., Marantz, L. B., McArthur, M. J., and Maxwell, M. H. (1969): Adsorbent based low volume recirculating dialysate. *Trans. Amer. Soc. Artif. Intern. Organs, 15*:347.

Gott, V. L., Whiffen, J. D., Koepke, D. E., Daggell, R. L., Boake, W. C., and Young, W. P. (1964): Techiques of applying a graphite-benzal-konium-heparin coating to various plastics and metals. *Trans. Amer. Soc. Artif. Intern. Organs,* 10:213.

Gottschalk, A. (1957): Virus enzymes and virus templates. *Physiol. Rev.,* 37:66.

Gottschalk, A. (1960): A correlation between composition, structure, shape and function of a salivary mucoprotein. *Nature,* 186:949.

Gottschalk, A. (1967): Report on artificial kidney, National Institute of Health.

Green, B. K., and Schneidcher, L. (1957): Microcapsules with hydrophobic contents. U.S. Pat. 2,800,457; U.S. Pat. 2,800,458.

Gregor, H. P., and Sollner, K. (1946): Improved methods of preparation of "permslective" collodion membranes combining extreme ionic selectivity with high permeability. *J. of Physical Chemistry,* 50:53.

Grode, G. A., Falb, R. D., Takahashi, M. T., and Leininger, R. I. (1968): Nonthrombogenic plastics. *Symposia on Bio-Chemical Engineering materials,* Part I, Philadelphia.

Grodins, F. S., Ivy, A. C., Adler, H. F. (1943): Intravenous administration of oxygen. *J. Lab. Clin. Med.,* 28:1009.

Gryszkiewicz, J. (1971): Insoluble Enzymes. Folia Biologica, 19:119.

Gutte, B., and Merrifield, R. B. (1969): The total synthesis of an enzyme with ribonuclease A activity. *J. Amer. Chem. Soc.,* 91:501.

Hagstam, K. E., Larsson, L. E., and Thysell, H. (1966): Experimental studies on charcoal hemoperfusion in phenobarbital intoxication and uremia including histological findings. *Acta. Med. Scand,* 180:593.

Halpern, B. N., Benacerraf, B., Biozzi, G., and Stiffel, C. (1958): *Physiopathology of the Reticulo-endothelial System.* Backwell, Oxford, p. 52.

Hampers, C. L., and Schupak, E. (1967): *Long-term Hemodialysis.* Grune & Stratton, New York.

Harris, J. W. (1963): *The Red Cell.* Harvard University Press, Cambridge, Mass.

Haskell, C. M., Canellos, G. P., Levanthal, B. G., Carbone, P. P., Block, J. B., Serpick, A. A., and Selaway, O. S. (1969): L-asparaginase: therapeutic and toxic effects in patients with neoplastic disease. *New Eng. J. Med.,* 281:1028.

Herbig, J. A. (1968): *Encyclopedia of Polymer Technology.*

Ho, P. P. K., Frank, B. H., and Burck, P. J. (1969): Crystalline L-asparaginase from *Escherichia coli* B. *Science,* 165:510.

Holland, R. A., and Forster, R. E. (1966): The effect of size of red cells on the kinetics of their oxygen uptake. *J. Gen. Physiol.,* 49:727.

Hori, M., and Toyoda, T. (1962): Investigations into blood replacement. *Sogo Igaku* (Tokyo), 19:191.

Hsia, David Yi-Yung (1966): *Inborn Errors of Metabolism.* The Year Book Publishers, Chicago.

Hutson, D. G., Rutledge, R. R., and Gollan, F. (1968): In Norman, J. C., et al. (Eds.), *Organ Perfusion and Preservation.* Appleton-Century-Crofts, New York.

Jandl, J. H., Simmons, R. L., and Castle, W. B. (1961): Red cell filtration and the pathogenesis of certain hemolytic anemia. *Blood, 18*:133.

Jay, A. W. L. (1967): Mechanical properties of an artificial cell membrane. M.Sc. Thesis, University of New Brunswick, Fredericton.

Jay, A. W. L., and Burton, A. C. (1969): Direct measurement of potential difference across the human red blood cell membrane. *Biophys. J., 9*:115.

Jay, A. W. L., and Edwards, M. A. (1968): Mechanical properties of semipermeable microcapsules. *Canad. J. Physiol. Pharmacol., 46*:731.

Jay, A. W. L., and Sivertz, K. S. (1969): Membrane resistance of semipermeable microcapsules. *J. Biomed. Mat. Res., 3*:577.

Johnson, L. (1967): Semipermeable microcapsules: biological effects of surface properties. M.Sc. Thesis, McGill University, Montreal.

Kedem, O., and Katchalsky, A. (1958): Thermodynamic analysis of the permeability of biological membranes to non-electrolytes. *Biochim. Biophys. Acta., 27*:229.

Kerr, D. N. S. (1967-1970): *Proceedings of the European Dialysis and Transplant Association,* Excerpta Medica Foundation.

Kidd, J. G. (1953): Regression of transplanted lymphoma induced *in-vivo* by means of normal guinea pig serum; I. Course of transplanted cancers of various kinds in mice and rat given guinea pig serum, horse serum or rabbit serum. *J. Exp. Med., 98*:565.

Kiil, F. (1960): Development of a parallel-flow artificial kidney in plastics. *Acta. Chir. Scand., 253*:142.

Kitajima, M., Miyano, S., and Kondo, A. (1969): Studies on enzyme-containing microcapsules. *J. Chem. Soc. Japan (Kogyo Kagaku Zasshi), 72*:493.

Koishi, M., Fukuhara, N., and Kondo, T. (1968): Preparation of polyphthalamide microcapsules. *Chem. Pharm. Bull., 17*:804.

Koishi, M., Fukuhara, N., and Kondo, T. (1969): Preparation and some properties of sulfonated polyphthalamide microcapsules. *Canad. J. of Chemistry, 47*:3447.

Kolff, W. J. (1970): Artificial Organs in the seventies. *Trans. Amer. Soc. Artif. Intern. Organs, 16*:534.

Kolff, W. J., and Berk, H. T. J. (1944): The artificial kidney: A dialyzer with a great area. *Acta. Med. Scand., 117*:121.

Kolff, W. J., and Watschinger, B. (1956): Results in patients treated with the coil kidney (disposable dialyzing unit). *J. Lab. Clin. Med., 47*:969.

Kondo, A. (1968): Personal communication.

Korn, E. D. (1968): Structure and function of the plasma membrane. *J. Gen. Physiol., 52*:2575.

Kusserow, B. K., Machanic, B., Collins, F. M., Jr., and Clapp, J. F. (1965): Changes observed in blood corpuscles after prolonged perfusions with two types of blood pumps. *Trans. Amer. Soc. Artif. Intern. Organs, 11:* 122.

Lawson, L. J., Crawford, N., Edwards, P. D., and Blainey, D. (1966): Platelet destruction and serotonin release during hemodialysis. *Proceedings of the European Dialysis and Transplant Association, 2:*63.

Leininger, R. I., Cooper, C. W., Epstein, M .M., Falb, R. D., and Grode, F. A. (1966): Nonthrombogenic plastic surfaces. *Proc. Nat. Acad. Sci., 56:*1828.

Lenard, J., and Singer, S. J. (1966): Protein conformation in cell membrane preparations as studied by optical rotary dispersion and circular dichroism. *Proc. Nat. Acad. Sci., 56:*1828.

Leonard, E. F. (1966): Chemical Engineering in Medicine. *Eng. Progr. Symp. Series, 66:*62.

Levine, S. N., and LaCourse, W. C. (1967): Materials and design consideration for a compact artificial kidney. *J. Biomed. Mat. Res., 1:*275.

Linn, B. S., Canaday, W., Berton, M., and Gollan, F. (1967): Kidney preservation by perfusion with organic liquids. *Surg. Forum, 18:*278.

Lucy, J. A. (1964): Globular lipid micelles and cell membranes. *J. Theor. Biol., 7:*360.

Luzzi, L. A. (1970): Preparation and evaluation of the prolonged release properties of nylon microcapsules, *J. Pharm. Sci., 59:*338.

Luzzi, L. A. (1970): Microencapsulation. *J. Pharm. Sci., 59:*1367.

Lyman, D. J., Klein, K. G., Brash, J. J., Fritzinger, B. K., and Andrade, J. D. (1970): Protein coated surfaces. Thromb. Diath. Haem. (In press).

Martel, L., Rotteveel, J., Snider, M. T., and Galletti, P. M. (1970): An extracorporeal chemical reactor. *Trans. Amer. Soc. Artif. Intern. Organs, 16:*504.

Mashburn, L. T., and Wriston, J. C. (1964): Tumor inhibitory effect of L-asparaginase from *Escherichia coli. Arch. Biochem. Biophys., 105:*450.

Maurer, F. W., Warren, M. F., and Drinker, C. K. (1940): The composition of mammalian pericardial and peritoneal fluids. Studies of their protein and chloride contents and the passage of foreign substances from the bloodstream into these fluids. *Amer. J. Physiol., 129:*635.

Merrifield, R. B. (1965): Automated synthesis of peptides. *Science, 150:* 178.

Merrill, E. W., Salzman, E. W., Lipps, B. J., Jr., Gilliland, E. R., Austen, W. G., and Joison, J. (1966): Antithrombogenic cellulose membranes for blood dialysis. *Trans. Amer. Soc. Artif. Intern. Organs, 12:*139.

Merrill, J. P. (1965): *Treatment of Renal Failure.* Grune & Stratton, New York.

Minoshima, T. (1965): *Nippon Seirigaku Zasshi, 27:*469.

Mirkovitch, V., Beck, R. E., Andrus, P. G., and Reininger, Q. Y., (1964):

the zeta potentials and blood compatibility characterics of some selected solids. *J. Surg. Res., 4*:395.

Mitchison, J. M., and Swann, M. M. (1954): The mechanical properties of the cell surface; I. The cell elastimeter. *J. Exp. Biol., 31*:443.

Morgan, P. W. (1965): *Condensation Polymers.* John Wiley & Sons, New York.

Morgan, P. W., and Kwolek, S. L. (1959): Interfacial polymerisation. *J. Polymer Sci., 40*:299.

Mosbach, K. (1970): Matrix-bound enzymes. I. The use of different acrylic copolymers as matrices. *Acta Chem. Scand. 24*:2084.

Mosbach, K., and Mattiasson, B. (1970): Matrix-bound enzymes. II. Studies on a matrix-bound two-enzyme-system. *Acta Chem. Scand., 24*:2093.

Mueller, P., Rudin, D. O., Tien, H. T., and Westcott, W. C. (1962): Bimolecular lipid membrane. *Nature, 194*:979.

Mueller, P., and Rudin, D. O. (1968): Resting and action potentials in experimental bimolecular lipid membranes. *J. Theor. Biol., 18*:222.

Muirhead, E. E., and Reid, A. F. (1948): A resin artificial kidney. *J. Lab. Clin. Med., 33*:841.

Nack, H. (1970): Microencapsulation. *J. Soc. Cosmetic Chemists, 21*:85.

Nawab, M. A., and Mason, S. G. (1958): Electrical emulsification. *J. Colloid Sci., 13*:179.

Nealon, T. F., Jr., and Ching, N. (1962): An extracorporeal device to lower blood ammonia levels in hepatic coma. *Trans. Amer. Soc. Artfi. Intern. Organs, 8*:226.

Nealon, T. F., Sugarman, H., Shea, W., and Fleegar, E. (1966): An extracorporeal device to treat barbiturate poisoning. Use of anion-exchange resins in dogs. *J.A.M.A., 197*:158.

Nose, Y. (1969): *The Artificial Kidney.* C. V. Mosby, St. Louis.

Nose, Y., Kon, T., Weber, D., Mrava, G., Malchesky, P., MacDermott, H., William, C., Lewis, L., Hoffman, G., Willis, C., Deadhar, S., Harris, G., and Anderson, R. (1970): Physiological effects of intravascular fluorocarbon liquids. *Fed. Proc., 29*:1789.

Pagano, R., and Thompson, T. E. (1967): Spherical lipid bilayer membranes. *Biochem. Biophys. Acta., 144*:666-669.

Pagano, R., and Thompson, T. E. (1968): Spherical lipid bilayer membranes. *J. Molec. Biol., 38*.

Pallotta, A. J., and Koppanhi, T. (1960): The use of ion exchange resins in the treatment of phenobarbital intoxication in dogs. *J. Pharmacol. Exp. Ther., 128*:318.

Pappenheimer, J. R. (1953): Passage of molecules through capillary walls. *Physiol. Rev., 33*:387.

Pappenheimer, J. R., Renkin, E. M., and Borrero, L. M. (1951): Filtration, diffusion and molecular sieving through peripheral capillary membranes; contribution to pore theory of capillary permeability. *Amer. J. Physiol., 167*:13.

References   191

Pauline, L. (1956): In Gaebler, O. H. (Ed.), *Enzymes*. Academic Press, New York.

Philpot, F. J., and Philpot, J.St.L. (1936): A modified colorimetric estimation of carbonic anhydrase. *Biochem. J., 30*:2191.

Poznansky, M. J. (1970): Microencapsulated catalase for enzyme replacement in acatalasemic mice. Ph.D. Thesis, McGill University, Montreal.

Poznansky, M. J., and Chang, T. M. S. (1969): Kinetics of semipermeable microcapsules containing catalase. *Proc. Canad. Fed. Biol. Sci., 28*:87.

Quinton, W., Dillard, D., and Scribner, B. H. (1960): Cannulation of blood vessels for prolonged hemodialysis. *Trans. Amer. Soc. Artif. Intern. Organs, 6*:104.

Rambourg, A., Neutra, M., and Leblond, C. P. (1966): Presence of a "cell coat" rich in carbohydrate at the surface of cells in the rat. *Anat. Rec., 154*:41.

Rand, R. P., and Burton, A. C. (1963): Area and volume changes in hemolysis of single erythrocyte. *J. Cell Comp. Physiol., 61*:245.

Regan, J. D., Vodopick, H., and Takeda, S. (1966): Serine requirement in leukemic and normal blood cells. *Science, 163*:1452.

Ring, G. C., Blum, A. S., and Kurtbatov, T. (1961): Size of microspheres passing through pulmonary circuit in the dog. *Amer. J. Physiol., 200*:1191.

Robertson, J. D. (1960): The molecular structure and contact relationships of cell membranes. *Progr. Biophys. Biophys. Chem., 10*:343.

Rosenbaum, J. L., Onesti, G., and Brest, A. N. (1962): The rapid removal of cations from dogs with an ion-exchange column. *J.A.M.A. 180*:762.

Rosenbaum, J. L., Winsten, S., Kramer, M., Moros, J., and Raja, R. (1970): *Trans. Amer. Soc. Artif. Intern. Organs, 16*:134.

Rosenthal, A., and Chang, T. M. S. (1971): The effect of valinomycin on the movement of rubidium across lipid coated semipermeable microcapsules. *Proc. Canad. Fed. Biol. Soc., 14*:44.

Rubin, A. L., Riggio, R. R., Nachman, R. L., Schwartz, G. H., Miyata, T., and Stanzel, K. H. (1968): Collagen materials in dialysis and implantation. *Trans. Amer. Soc. Artif. Intern. Organs, 14*:169.

Russell, P. S., and Monaco, A. P. (1964): The biology of tissue transplantation. *New Eng. J. Med., 271*: 502-510.

Sachtleben, P., and Ruhenstroth-Bauer, G. (1961): Agglutination and the electrical surface potential of red blood cells. *Nature, 192*:982.

Sawyer, P. N., and Pate, J. W. (1953): Bio-electric phenomena as etoilogic factor in intravascular thrombosis. *Amer. J. Physiol., 175*:103.

Schechter, D. C., Nealon, T. F., Jr., and Gibbon, J. H., Jr. (1958): An ion exchange resin artificial kidney. *Surg. Forum, 9*:1958.

Schreinder, G. E., Ed. (1955-1970): Development of Artificial Organs. *Trans. Amer. Soc. Artif. Intern. Organs*, Georgetown University Printing Department, Washington, D.C.

Schreiner, G. E., and Maher, J. F. (1961): *Uremia*. Thomas, Springfield.

Schulman, J. H., and Rideal, E. K. (1937): Molecular interaction in mono-layers. *Proc. Roy. Soc. Biol.*, *122*:29.

Scriver, C. R. (1969): Treatment of inherited disease: realized and po-tential. *Med. Clin. N. Amer.*, *53*:941.

Seaman, G. V. F., and Swank, R. L. (1963): Surface characteristics of dog erythrocytes and some lipid models. *J. Physiol.*, *168*:118.

Sekiguchi, W., and Kondo, A. (1966): Studies on microencapsulated hemo-globin. *J. Jap. Soc. Blood Transfusion*, *13*:153.

Seufert, W. D. (1970): Model membranes: spherical shells bounded by one bimolecular layer of phospholipids. *Biophysik.*, *7*:60.

Sharp, W. V., Gardner, D. L., and Andressen, G. L. (1965): Adaptation of elastic materials for small vessel replacement. *Trans. Amer. Soc. Artif. Intern. Organs*, *11*:336.

Shiba, M., Tomioka, S., Koishi, M., and Kondo, T. (1970): Studies on microcapsules containing aqueous protein solution. *Chem. Pharm. Bull.*, *18*:803.

Shigeri, Y., Koishi, M., Kondo, T., Shiba, M., and Tomioka, S. (1970): Studies on microcapsules: variations in microcapsule size. *Canad. J. Chem.*, *48*:2047.

Shigeri, Y., and Kondo, T. (1969): Permeability of polyurethane micro-capsule membranes. *Chem. Pharm. Bull.*, *17*:1073.

Silman, J. H., and Katchalsky, E. (1966): Insolubilized enzymes. *Ann. Rev. Biochem.*, *35*:873.

Sjostrand, F. S. (1963): A new repeat structural element of mitochondrial and certain cytoplasmic membranes. *Nature*, *199*:1262.

Skeggs, L. T., Jr., and Leonard, J. R. (1948): Studies of an artificial kid-ney: preliminary results with a new type of continuous dialyzer. *Science*, *108*:212.

Sloviter, H. A. (1970): Erythrocyte substitutes. *Med. Clin. N. Amer. 54*: 787.

Sloviter, H. A., and Kaminoto, T. (1967): Erythrocyte substitute for per-fusion of brain. *Nature*, *216*:458.

Sloviter, H. A., Petkovic, M., Ogoshi, S., and Yamada, H. (1969): Dispersed fluorochemicals as substitutes for erythrocytes in intact animals. *J. Appl. Physiol.*, *27*:666.

Sloviter, H. A., Yamada, H., and Ogoski, S. (1970): Some effects of intra-venously administered dispersed fluorochemicals in animals. *Fed. Proc.*, *29*:1755.

Sollner, K. (1958): Ion selective membranes. *Svensk Kemisk Tidskrift*, 6-7, 268.

Solomon, A. K. (1961): In Klinzeller, A., and Kotyk, A. (Eds.): *Membrane Transport and Metabolism*. Academic Press, New York.

Solomon, A. K. (1969): Characterization of biological membranes by equivalent pores. *J. Gen. Physiol.*, *51*:335.

Sparks, R. E., Blaney, J. L., and Lindan, O. (1966): Adsorption of nitrogenous waste metabolites from artificial kidney dialyzing fluid. *Chem. Eng. Progr. Symp. Series, 62*:2.

Sparks, R. E., Salemme, R. M., Meier, P. M., Litt, M. H., and Lindan, O. (1969): Removal of waste metabolites in uremia by microencapsulation reactants. *Trans. Amer. Soc. Artif. Intern. Organs, 15*:353.

Sparks, R. E., Lindan, O., Mason, N. S., Litt, M. H., and Meier, P. M. (1971): Removal of uremic waste metabolites from the gastrointestinal tract by encapsulated carbon and oxidized starch. *Trans. Amer. Soc. Artif. Intern. Organs* (in press).

Stanfield, R. (1968): Significance of dialysis against enzymes to the specific therapy of cancer and genetic deficiency diseases. *Nature, 220*:1321.

Stavermann, A. J. (1951): The theory of measurement of osmotic pressure. *Rec. Trav. Chim., 70*:344.

Stein, W. D. (1967): The movement of Molecules Across Cell Membranes, *Academic Press,* New York.

Stewart, R. D., Baretta, E. D., Cerny, J. C., and Mahon, H. (1966): A capillary artificial kidney. *Invest. Urol., 3*:614.

Stone, H. B., and Kennedy, W. J. (1964): Survival of heterologous mammalian transplants: a third report: *Ann. Surg., 159*:645.

Suzuki, S., Kondo, A., and Mason, S. G. (1968): Studies on microcapsules. *Chem. Pharm. Bull., 16*:1629.

Tobias, J. M. (1964): A chemically specified molecular mechanism underlying excitation in nerve: a hypothesis. *Nature, 203*:13.

Toyoda, T. (1965): Blood replacement. *Kagaku, 36*:7.

Toyoda, T. (1966): Fundamental studies on artificial blood. *J. Surg. Soc. Jap., 67*:36.

Trinder, L., Verosky, M., Habif, D. V., and Nahas, G. G. (1970): Perfusion of isolated liver with fluorocarbon emulsions. *Fed. Proc., 29*:1778.

Van Slyke, D. D., and Archibald, R. M. (1944): Manometric, titrimetric and colonmetric methods for measurement of urease activity. *J. Biol. Chem., 154*:623.

Wang, J. H. (1958): Haemoglobin studies; II. A synthetic material with haemoglobin-like property. *J. Amer. Chem. Soc., 80*:3168.

Whiffen, J. D., and Gott, V. L. (1965): In-vivo adsorption of heparin by graphite-benzalkonium intravascular surfaces. *Surgery, 121*:287.

Wilkins, D. J., and Meyers, P. (1966): Studies on the relationship between the electrophoretic properties of colloids and their blood clearance and organ distribution in the rat. *Brit. J. Exp. Path., 47*:568.

Yatzidis, H. (1964): A convenient hemoperfusion micro-apparatus over charcoal for the treatment of endogenous and exogenous intoxications. *Proceedings of the European Dialysis and Transplant Association, 1*:83.

Ziegler, E. E. (1941): The intravenous administration of oxygen. *J. Lab. Clin. Med., 27*:223.

INDEX

195